Johannes Amon

Mining The Genomes Of Actinomycetes

Identification Of Metabolic Pathways And Regulatory Networks

Südwestdeutscher Verlag für Hochschulschriften

Impressum / Imprint
Bibliografische Information der Deutschen Nationalbibliothek: Die Deutsche Nationalbibliothek verzeichnet diese Publikation in der Deutschen Nationalbibliografie; detaillierte bibliografische Daten sind im Internet über http://dnb.d-nb.de abrufbar.
Alle in diesem Buch genannten Marken und Produktnamen unterliegen warenzeichen-, marken- oder patentrechtlichem Schutz bzw. sind Warenzeichen oder eingetragene Warenzeichen der jeweiligen Inhaber. Die Wiedergabe von Marken, Produktnamen, Gebrauchsnamen, Handelsnamen, Warenbezeichnungen u.s.w. in diesem Werk berechtigt auch ohne besondere Kennzeichnung nicht zu der Annahme, dass solche Namen im Sinne der Warenzeichen- und Markenschutzgesetzgebung als frei zu betrachten wären und daher von jedermann benutzt werden dürften.

Bibliographic information published by the Deutsche Nationalbibliothek: The Deutsche Nationalbibliothek lists this publication in the Deutsche Nationalbibliografie; detailed bibliographic data are available in the Internet at http://dnb.d-nb.de.
Any brand names and product names mentioned in this book are subject to trademark, brand or patent protection and are trademarks or registered trademarks of their respective holders. The use of brand names, product names, common names, trade names, product descriptions etc. even without a particular marking in this works is in no way to be construed to mean that such names may be regarded as unrestricted in respect of trademark and brand protection legislation and could thus be used by anyone.

Verlag / Publisher:
Südwestdeutscher Verlag für Hochschulschriften
ist ein Imprint der / is a trademark of
OmniScriptum GmbH & Co. KG
Heinrich-Böcking-Str. 6-8, 66121 Saarbrücken, Deutschland / Germany
Email: info@svh-verlag.de

Herstellung: siehe letzte Seite /
Printed at: see last page
ISBN: 978-3-8381-1740-9

Zugl. / Approved by: Erlangen, FAU, Diss., 2010

Copyright © 2010 OmniScriptum GmbH & Co. KG
Alle Rechte vorbehalten. / All rights reserved. Saarbrücken 2010

For my parents

I may not have gone where I intended to go,

but I think I have ended up where I needed to be.

Douglas Adams, British humorist & science fiction novelist (1952 - 2001)

This thesis contains publications of the following publishers that are reproduced with permission:

ASM Journals, Washington D.C., USA

Horizon Scientific Press/Caister Academic Press, Norfolk, UK

S. Karger AG, Basel, Switzerland

Table of Contents

1 Introduction 1

1.1 Gram-positive bacteria: the order Actinomycetales 1

1.2 Actinomycetes: The era of genomics 7

1.3 The basics of nitrogen metabolism and regulation 10

1.4 The basics of carbohydrate uptake and regulation 13

1.5 Proteolysis: An alternative route for carbohydrate and nitrogen sources and its role in pathogenicity 15

1.6 Aim 18

2 Results & Discussion 19

2.1 Regulation of nitrogen metabolism in *Mycobacterium smegmatis* 19

2.1.1 Nitrogen-dependent expression of ammonium transport and assimilation proteins depends on the OmpR-type regulator GlnR 19

2.1.2 The role and function of AmtR in *M. smegmatis* and *S. avermitilis* 20

2.1.3 The search for the GlnR interaction partner 22

2.2 Comparative genomic analysis of nitrogen metabolism and control in mycobacteria 24

2.3 The glucose permease and glucose kinase of *M. smegmatis* 26

2.4 Carbohydrate transport systems of mycobacteria 30

2.5 Carbohydrate transport systems of *Bifidobacterium longum* 32

2.6 Comparative genomic analysis of the proteolytic potential in corynebacteria 34

3 Summary / Zusammenfassung 36

3.1 Summary 36

		3.2	Zusammenfassung	37
4	**References**			**38**
5	**Own publications**			**49**
6	**Appendix**			**50**
	6.1	Full species names		50
	6.2	Publications		52

1 Introduction

1.1 Gram-positive bacteria: the order Actinomycetales

In 1884, the Danish scientist Hans-Christian Gram (1853 – 1938) published a technique of staining bacteria to distinguish two species with similar clinical symptoms, namely *Streptococcus pneumoniae* and *Klebsiella pneumoniae*. This staining method, which is until today one of the most basic and widely used laboratory techniques for the characterization of a bacterial organism, is based on the chemical and physical properties of the cell wall and allows to differentiate bacterial species into two large groups, namely the Gram-negative and the Gram-positive bacteria. While the technique of Gram-staining for exact identification purposes in modern environmental or molecular microbiology is now mostly superseded by advanced techniques such as genetic sequencing, Gram-stains remain a reliable and fast detection method in medical laboratories when a bacterial infection is suspected and the patient's treatment (e.g. the appropriate antibiotics choice) has to be determined.

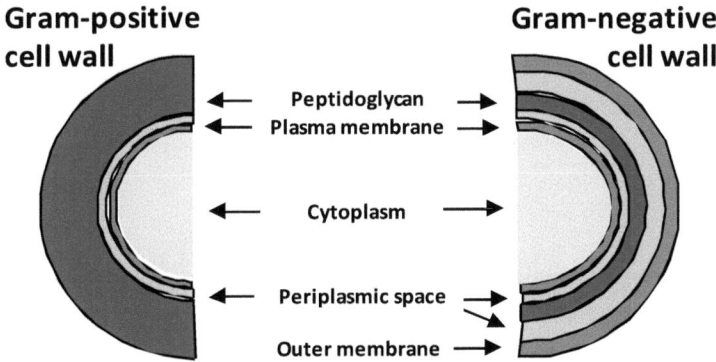

Figure 1: Gram-positive and -negative cell wall structure. Note that Gram-positive organisms feature a much thicker peptidoglycan layer containing lipoteichoic acid (LTA), while Gram-negative bacteria have an outer membrane containing lipopolysaccharide (LPS).

Apart from the shortcomings of this technique such as some organisms being Gram-variable or not susceptible to staining at all, the differentiation of bacteria into Gram-negative and Gram-positive organisms proves itself valid even today in the era of genomics where high-throughput sequencing allows the rapid creation and detailed phylogenetic analysis of whole genome sequences. This is mostly owed to the fact that Gram-staining exploits one of the most fundamental differences between bacterial organisms, namely the occurrence of a thick peptidoglycan cell wall for Gram-positive organisms in comparison to the thin cell wall and outer membrane of Gram-negative bacteria (Fig. 1). Gram-positive bacteria lack the outer membrane and associated lipopolysaccharide (LPS) that is present in Gram-negative organisms. In Gram-positive bacteria, the peptidoglycan layer is thicker and contains teichoic acids. An overview of bacterial phylogeny is given in Fig. 2.

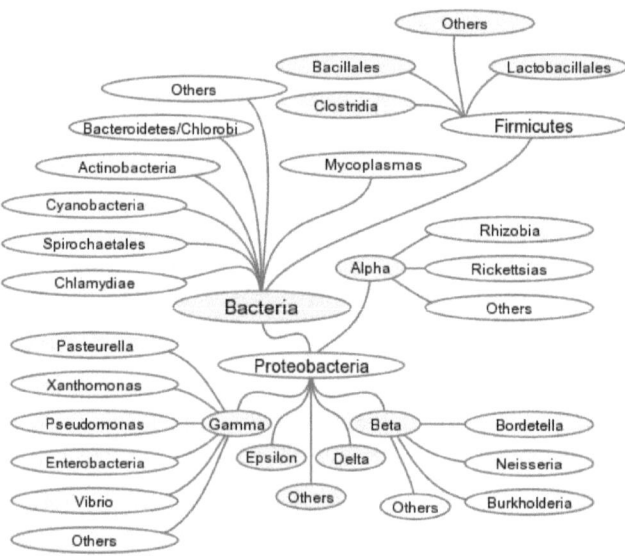

Figure 2: Overview of bacterial phylogeny. Note that within all given phyla, only the Actinobacteria and Firmicutes are Gram-positive. Source: NCBI Bacterial Genomes Tree BLAST group, National Institutes Of Health, Bethesda, MD, USA.

Focussing on the group of Gram-positive organisms, another subdivision was created which is based on the genomic content of the nucleotides guanine and cytosine; this resulted in two major groups, namely the phyla of low G+C Firmicutes (e.g. bacilli, clostridia, cocci) and high G+C Actinobacteria (e.g. streptomycetes, corynebacteria, mycobacteria, bifidobacteria; for review on phylogenetics, see Ventura *et al.*, 2007b). Despite sharing the thick peptidoglycan cell wall, the two phyla show only distant relationship on a phylogenetic level, which is also reflected on 16S rRNA analyses (Fig. 3).

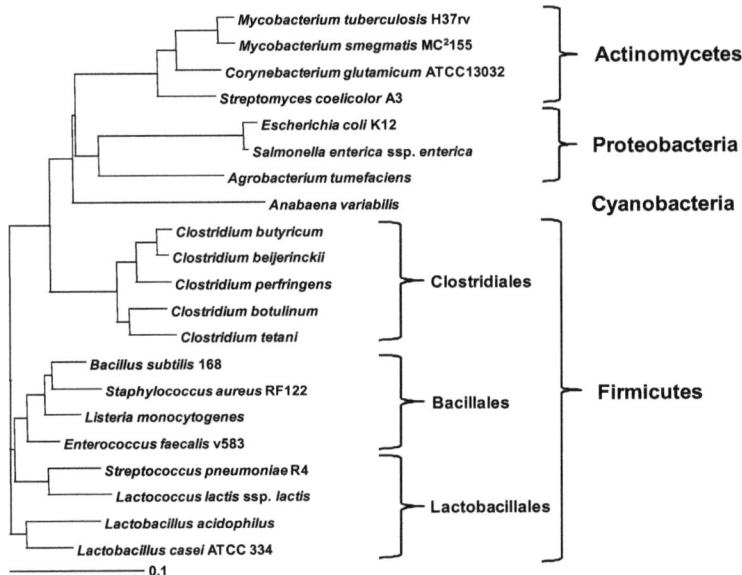

Figure 3: Phylogenetic tree of selected bacteria based on 1,500 nucleotides of 16S rRNA. Scale bar, substitutions of 5 nucleotides. Note that the Proteobacteria and Cyanobacteria are Gram-negative, while the Actinobacteria and Firmicutes are Gram-positive organisms. Phyla and orders are indicated.

Within the phylum Actinobacteria, the streptomycetes take a special role in many respects. The most distinguishable feature is their capability of morphological differentiation: most of them grow vegetatively in a mycelium-like filamentous fashion and are able to form aerial branches which in turn produce spores as a result of nutrient depletion (Fig. 4; Chater, 1993). Sporulation of aerial hyphae is connected to the other distinct feature of streptomycetes, the activation of secondary metabolism and production of antibiotics (Hopwood, 1988). This capability has turned streptomycetes into the primary antibiotics-producing organisms for the pharmaceutical industry (Berdy, 2005).

Introduction

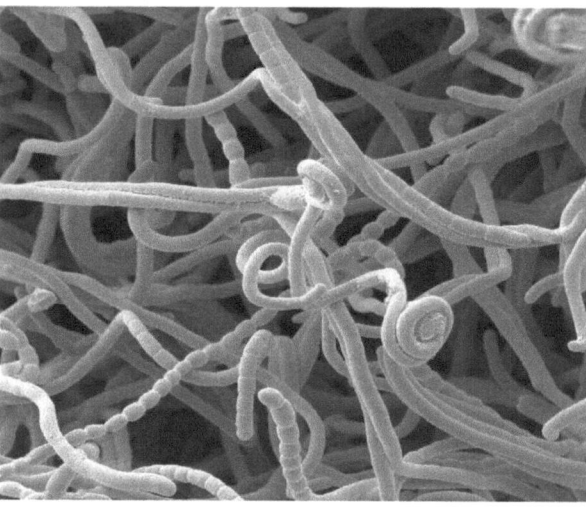

Figure 4: Scanning electron micrograph of the aerial hyphae and spores of *Streptomyces coelicolor*. Source: Dr. Paul Hoskisson, University of Strathclyde, Glasgow, UK.

Other human related, industrial relevant Actinobacteria belong to the deep-branching family of *Bifidobacteriaceae*. Isolated from ecological niches such as the intestine, the oral cavity and food, they show a far simpler morphology than the streptomycetes (Ventura *et al.*, 2007a). Of special interest are the gastrointestinal tract (GIT) inhabitants due to their probiotic properties. GIT species such as *Bifidobacterium breve* or *Bifidobacterium longum* (biotype *longum* and biotype *infantis*) are able to uptake and ferment a large variety of di-, tri- and oligosaccharides, especially those that are not digested by their hosts (Ventura *et al.*, 2007a). While mammalians commonly consume plant-based foods which have a high rate of complex polysaccharides that contain sugars such as glucose, fructose, or arabinose, mammalian genomes apparently lack coding sequences for enzymes that are capable of degrading most of these glycans. GIT bifidobacteria on the other side are equipped with a multitude of glycosyl hydrolases that are predicted to be involved in the degradation of these higher order poly- and oligosaccharides (Schell *et al.*, 2002).

Inside the Actinobacteria, the genera *Corynebacterium*, *Mycobacterium* and *Nocardia* form a monophyletic taxon, the so-called CMN group (Embley & Stackebrandt, 1994). These bacteria are classified as Gram-positive bacteria, but they have features of both Gram-positive and Gram-negative organisms. Members of the CMN group share an unusual cell envelope composition, characterized by the presence of a wax-like cell envelope containing mycolic acids (Fig. 6). Thus, the production of mycolic acids seems to be a genuine trait unique to this phylogenetic group (Embley & Stackebrandt, 1994).

Figure 5: Scanning electron micrograph of cells of *Corynebacterium glutamicum* immobilized on carrier material. Source: Forschungszentrum Jülich.

The genus *Corynebacterium* comprises human relevant species such as *Corynebacterium jeikeium*, a lipid-requiring commensal organism of the human skin flora that was also associated with nosocomial infections (Coyle & Lipsky, 1990), and one of the most notorious human pathogens, *Corynebacterium diphtheriae*, the causative agent of diphtheria (Hadfield *et al.*, 2000). The disease is caused by exotoxin-producing *C. diphtheriae* strains and characterized by the formation of a pharyngal pseudomembrane. The diphtheria toxin, which is encoded by corynephages, belongs to the family of AB toxins that inhibit protein synthesis and finally kill susceptible host cells (Holmes, 2000). Far more beneficial members of this genus are *Corynebacterium glutamicum*, where genetically modified high-performance strains are widely used in the industrial production of amino acids such as L-glutamic acid, L-lysine, L-serine and L-threonine (Leuchtenberger *et al.*, 2005), and the close relative *Corynebacterium efficiens*, strains of which are able to grow at higher temperatures than *C. glutamicum*; this makes *C. efficiens* an interesting target for biotechnological use in fermenters and as a supplemental amino acid-producing organism for *C. glutamicum* as these strains might reduce the need for cooling during the process of fermentation (Fudou *et al.*, 2002).

Introduction

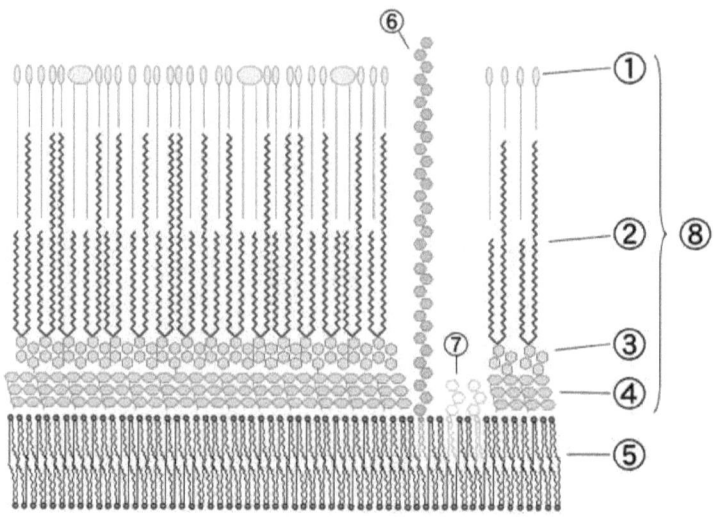

Figure 6: Schematic diagram of the mycobacterial cell wall. 1: outer lipids, 2: mycolic acids, 3: polysaccharides (arabinogalactan), 4: peptidoglycan, 5: plasma membrane, 6: lipoarabinomannan (LAM), 7: phosphatidylinositol mannoside, 8: cell wall skeleton. Figure used under license of the GNU Free Documentation License FDL 1.2.

Moving along in the CMN group, the genus *Mycobacterium* is divided between slow-growing and fast-growing organisms. The slow-growing members are mostly the pathogenic mycobacteria including the causative agents of tuberculosis and leprosy, which are both intracellular parasites. Not much is known about the exact virulence and pathogenicity of *Mycobacterium leprae*, also due to the fact that the only known animal model susceptible to leprosy is the nine-banded armadillo (Job, 2003). This not only complicates research, but also emphasizes the importance of one or more yet unknown genetic host factors that allow susceptibility to the leprosy bacillus (Ranque *et al.*, 2008). Also, being an obligate intracellular parasite, the genome of *M. leprae* shows the most serious case of gene loss and decay, a process termed "reductive evolution" (Gómez-Valero *et al.*, 2007). A more complete picture of pathogenicity and virulence factors is available for *Mycobacterium tuberculosis*, which is replicating within modified phagosomes of macrophages. In this location, the bacterial cells prevent formation of phagolysosomes and inhibit the macrophage responses to infection such as apoptosis and secretion of inflammatory cytokines (Flannagan *et al.*, 2009). These activities depend on cell wall constituents, such as plasma membrane

Introduction

lipoarabinomannans, which consist of phosphatidylinositol mannosides covalently linked to arabinogalactans that extend out from the cell surface (Figure 6). As cells of *M. tuberculosis* only divide every 15-20 hours, which seems to be a distinct feature also of other pathogenic *Mycobacterium* species such as *Mycobacterium bovis* (causing the cattle tuberculosis) or *Mycobacterium avium* ssp. *paratuberculosis* (causing Johne's disease in ruminants), the fast-growing non-pathogenic soil bacterium *Mycobacterium smegmatis* is often used in research as a model organism to study common features of mycobacterial biology such as metabolic and regulatory pathways or the genus-typical cell wall composition.

1.2   Actinomycetes: The era of genomics

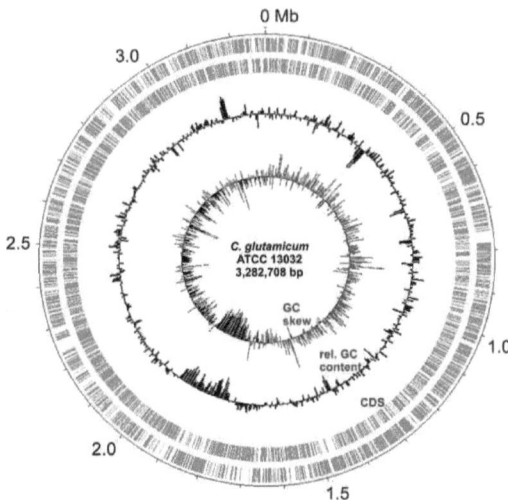

Figure 7: Circular representation of the *C. glutamicum* ATCC 13032 chromosome. The concentric circles denote (from outward to inward): coding sequences (CDS) transcribed clockwise and counter-clockwise, relative G/C content, and GC skew. For details, see Kalinowski *et al.*, 2003.

Many of the recent advances in biology would not have been possible without the availability of genome sequencing. Especially in the field of microbiology, the number of available whole-genome sequences of various species grew nearly exponentially over the last few years. At the time of writing (Feb. 2010), the servers at the National Center for Biotechnology Information (NCBI) contain the completed and annotated genomes of 1021 microbial

Introduction

organisms (ftp://ftp.ncbi.nih.gov/genomes/Bacteria/), with 2260 prokaryotic genome sequencing projects available as draft assembly or in progress (http://www.ncbi.nlm.nih.gov/genomes/lproks.cgi). This increase is mostly owed to the ground-breaking pyrosequencing method (Ronaghi *et al.*, 1998), which is commercially available in the large-scale DNA sequencing systems by 454 Life Sciences. The lengths of individual reads of DNA sequence are significantly shorter than with the classic chain termination method (Sanger sequencing; Sanger & Coulson, 1975), but this limitation is mostly redeemed by the high-throughput and the consequent manifold coverage that is available by this technique. These array-based pyrosequencing platforms are able to generate 400 million nucleotide data in a 10 hour run, thus best suited for genome sequencing and metagenomics applications, so that the processing of the raw sequence data is now the more time consuming process.

Prokaryotic DNA is especially well suited for sequencing since it has some advantages compared to eukaryotic DNA. The bacterial chromosome is usually organized circular (Fig. 7), so replication is possible without the need for telomeres; furthermore, the majority of bacteria possess only one chromosome, not counting movable DNA elements such as plasmids, and as such is of much smaller size than eukaryotic chromosomes. Prokaryotic organisms divide by a simple process called binary fission, where, after DNA replication and chromosome segregation, the cell simply divides in the middle, resulting in two identical daughter cells. The multiplication is usually asexually, so there is no need for a distinction between mitosis or meiosis, which in turn results in no cell phases or cell differentiation as in eukaryotes. Thus the prokaryotic chromosome is not compacted by histones into various states of aggregation but for the most part simply packed by supercoiling, also if recent advances reported some unanticipated structural and functional complexity (Thanbichler *et al.*, 2005).

Dependent on the computed coverage of the whole bacterial genome, the number of all obtained contigs (gapless long sequencing reads) and the number of non-overlapping contigs, the post-processing in form of additional Sanger sequencing to manually fill the gaps between two adjacent contigs and the bioinformatics assembly and annotation pipeline can very well pose the more time-consuming process. This process is often sped up by using so-called scaffold genomes: these are high-quality genomes of closely related organisms that were done using traditional sequencing methods and that have been carefully manually annotated. Examples of these scaffold genomes are the whole genome sequences of *Escherichia coli* K12

Introduction

(Blattner *et al.*, 1997) for proteobacterial genome projects; *Bacillus subtilis* 168 (Kunst *et al.*, 1997) for Gram-positive low GC genome projects; *M. tuberculosis* H37rv (Cole *et al.*, 1998) for mycobacterial genomes; and *C. glutamicum* ATCC 13032 (Fig. 7; Ikeda *et al.*, 2003; Kalinowski *et al.*, 2003) for corynebacterial genome projects. After the fully automated step of determining open reading frames (coding sequences, CDS) and other features like ribosomal RNA sequences for phylogeny analyses inside the whole genome sequence (Fig. 7), homology analyses with the protein products of the predicted reading frames have to be done to determine putative homologs from other closely related organisms. This process, which is also mostly automated, involves BLAST searches with databases for already described and highly similar protein products (Altschul & Lipman, 1990) as well as domain prediction analyses involving Hidden Markov Models (HMMs) with domain databases such as the PFAM (Finn *et al.*, 2010).

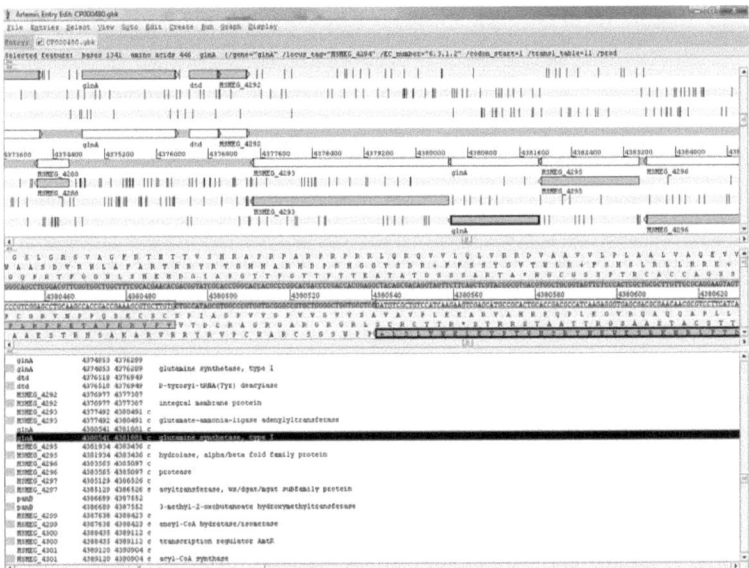

Figure 8: Sample screenshot of the *M. smegmatis* MC$_2$[155] genome, viewed with Artemis Version 11 (Carver et al., 2008). Automatically annotated CDS shown in the upper part, raw sequence with corresponding reading frames in the middle, gene names with location information and predicted products in the lower part.

While the automated annotation might give a rough overview of the coding potential of an organism's genome sequence, manual inspection and maintenance has to be done afterwards

Introduction

to find errors and mismatches in the CDS and their predicted products. These errors comprise, among others, wrong start and stop codons, false homology hits and incorrect gene names. Furthermore, advanced coding features in the intergenic spaces such as terminator structures, sRNAs or regulator binding sites are still not being annotated automatically, which requires manual screening of the sequences of interest with specialized bioinformatics tools and servers.

Summarizing, whole genome sequences of microorganisms represent an invaluable treasure of information to the microbiologist interested in the coding capability of a bacterium of interest, whether he is interested in pathogenicity and virulence features or the genetic potential for uptake and metabolism of various substrates, among many other possibilities.

1.3 The basics of nitrogen metabolism and regulation

For long time, nitrogen control in Gram-negative enterobacteria was the paradigm of nitrogen metabolism and regulation in bacteria. The pathways used for example by *E. coli* to assimilate ammonium depend on the concentration of this nitrogen source in the medium and are under strict control (for review, see Merrick & Edwards, 1995; Leigh & Dodsworth, 2007). During growth in ammonium-rich medium, this nitrogen source is primarily assimilated by glutamate dehydrogenase (GDH) while at ammonium concentrations below 1 mM, affinity of GDH to ammonium is too low and in this situation, the glutamine synthetase/glutamate synthase (GS/GOGAT) system takes over. The flux through GS is subject to regulation on the level of enzyme activity and expression of the GS-encoding *glnA* gene. In enteric bacteria, the activity of GS is regulated by adenylylation/deadenylylation, depending on the nitrogen availability. This modification is catalyzed by the bifunctional enzyme adenylyltransferase (ATase/GlnE), the activity of which is controlled by the signal transduction protein PII (GlnK/GlnB/NrgB). PII occupies a pivotal position in nitrogen regulatory networks as a sensory protein and signal transducer (for recent reviews, see Forchhammer 2007; 2008). The nitrogen status of the cell, as sensed by uridylyltransferase (GlnD), is signalled to PII by adjusting the degree of uridylylation of the latter. Native PII indicates a nitrogen-rich status, whereas PII-UMP flags a nitrogen-poor status of the cell. From the PII protein, the signal is transferred to ATase, which regulates GS activity, and to the NtrC protein, which is part of a two-component system involved in the transcriptional regulation of genes under nitrogen control (Weiss *et al.*, 2002). Two PII-type proteins are

Introduction

present in *E. coli*, the *glnB* (Bueno *et al.*, 1985; van Heeswijk *et al.*, 2009) and *glnK* gene product (van Heeswijk *et al.*, 1995; 1996). PII and GlnK form heterotrimers (Forchhammer *et al.*, 1999) that are proposed to be important for fine tuning the nitrogen regulatory cascade (van Heeswijk *et al.*, 2000). The complete *E. coli* signal transduction cascade was investigated in a reconstituted system (Jiang *et al.*, 1998a; 1998b; 1998c). These experiments verified that the *E. coli glnD* product, the uridylyltransferase, senses glutamine as a nitrogen signal. Glutamine inhibits uridylylation of PII by affecting the rate of UMP transfer and, in the presence of Mg^{2+}, stimulates deuridylylation of PII-UMP. Interaction of PII with NtrB (see below) is inhibited by uridylylation and, in addition, by binding of oxoglutarate to PII. Low levels of oxoglutarate stimulate the interaction of PII and NtrB, leading to an inhibition of kinase activity and an activation of phosphatase activity of NtrB.

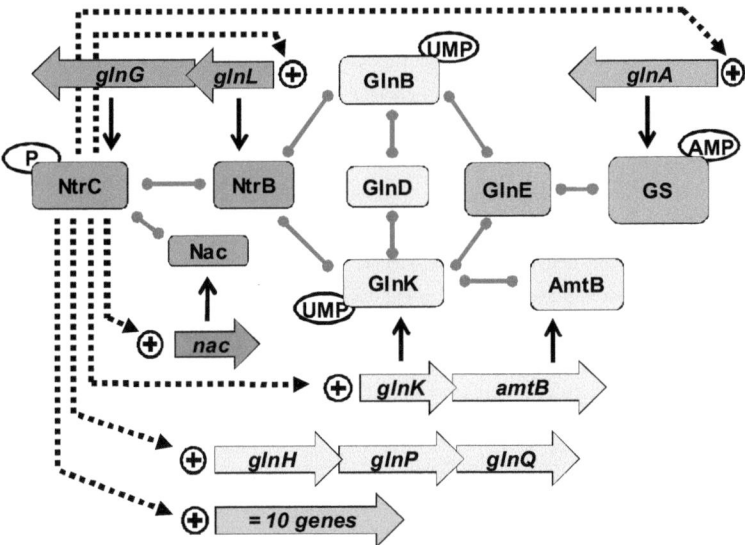

Figure 9: Transcriptional regulation of nitrogen metabolism in *E. coli*. Arrows indicate the relative gene length and organization on the chromosome with the respective gene name. Dotted black arrows illustrate the regulatory interactions of the respective transcription factor (orange) including its function in activating (plus) the target gene(s). Grey lines with circles indicate post-translational interactions. Blue, glutamine synthetase and GS ATase; yellow, transport systems and PII proteins; grey, further genes involved in nitrogen metabolism. *glnHPQ*: components of the *E.coli* glutamine permease. NtrC-dependent transcriptional activation is σ^{54}-dependent; NtrBC equivalent to NRII-NRI. For accessibility reasons, protein modifications such as de-/uridylylation, de-/adenylylation or de-/phosphorylation are not fully shown. Note that the genes *glnA-glnL-glnG* actually form an operon. For details about gene names and products, see text.

Introduction

The expression of *glnA* and several other genes involved in nitrogen metabolism in enterobacteria is controlled by the Ntr nitrogen regulation system (Fig. 9). This two-component signal transduction system activates transcription of σ^{54}-dependent promoters in response to nitrogen limitation. Phosphorylation of the response regulator NtrC by the NtrB kinase is controlled by the PII protein depending on the cellular glutamine concentration (via uridylylation/deuridylylation of PII by UTase) and depending on the cellular oxoglutarate concentration (Jiang & Ninfa, 2009a; 2009b; 2009c). High oxoglutarate concentrations prevent binding of PII to NtrB leading to a decrease of NtrB phosphatase and a stimulation of NtrB kinase activity. As a consequence, NtrC is phosphorylated and Ntr-regulated genes are expressed.

Subsequent work and comparative genomics analyses with a growing number of whole genome sequences showed a rather uniform regulation of nitrogen metabolism in the Gram-negative proteobacteria similar to the system described in detail above for *E. coli* (Leigh & Dodsworth, 2007). Outside this group, the situation is more complex and diverse. In the last decade, nitrogen control in several Gram-negative and especially Gram-positives was analyzed and it soon became clear that for virtually every phylum of the bacteria, different strategies of transcriptional regulation evolved, while the assimilatory enzymes (e.g. glutamine synthetase and glutamate synthase) as well as signal transduction proteins (adenylyltransferase, uridylyltransferase, PII) are widely conserved. For example, the Gram-negative cyanobacteria evolved the Crp-Fnr family regulator NtcA (Herrero *et al.*, 2001; Muro-Pastor *et al.*, 2005) for the transcriptional control of ammonium assimilation, whereas the post-translational signal transduction via the PII system - while sporting some unique features - is also for cyanobacteria highly conserved (Forchhammer, 2004; Osanai & Tanaka, 2007).

For the Actinobacteria, the regulation of nitrogen metabolism was already studied in detail for *Streptomyces coelicolor* (Reuther & Wohlleben, 2007; Tiffert *et al.*, 2008) and the genus *Corynebacterium* (Burkovski, 2007; Walter *et al.*, 2007; Hänßler & Burkovski, 2008), while not much was known about the regulatory networks of nitrogen metabolism in the mycobacteria, e.g. *M. smegmatis* and its pathogenic relatives *M. tuberculosis* and *M. leprae*.

Introduction

1.4 The basics of carbohydrate uptake and regulation

Initial work about sugar uptake and regulation focussed on the glucose-specific phosphoenolpyruvate-dependent phosphotransferase system (PTS II^{Glc}) in *E. coli* (Erni & Zanolari, 1986; for review, see Jahreis *et al.*, 2008). Since then, four more major uptake systems for glucose were discovered in *E. coli* (Fig. 10) including the mannose - PTS II^{Man} (Gutknecht *et al.*, 1999), the *N*-Acetylglucosamine - PTS II^{Nag} (Postma *et al.*, 1993), the galactose permease of the major facilitator superfamily (MFS) GalP (Weickert & Adhya, 1993; El Qaidi *et al.*, 2009), and the methyl-galactoside permease MglBAC (Death & Ferenci, 1994; Ferenci, 1996). The PTS is the major uptake system for carbohydrates in enteric bacteria and consists of conserved components: The cytoplasmic proteins Enzyme I (EI, encoded by *ptsI*) and the phosphocarrier protein HPr (encoded by *ptsH*) are at the beginning of the PTS-specific phosphorylation chain. After autophosphorylation of EI with the phosphate group of phosphoenolpyruvate (PEP), the phosphate group gets transferred from HPr to the hydrophilic membrane-associated proteins Enzymes IIA and IIB. The specific carbohydrate is then, by simultaneous phosphorylation from IIB, transported into the cell via the transmembrane permease components IIC and (optionally) IID. These proteins are present in the cell either as a single multidomain proteins or as several individual polypeptides, depending on the specific PTS and organism (Figure 10; Jahreis *et al.*, 2008). For example, while the *E. coli* Enzyme IIA^{Glc} is encoded by the *crr* gene and the Enzyme IICB^{Glc} is the product of the *ptsG* gene, the *B. longum ptsG* gene encodes the complete Enzyme IIABC^{Glc}.

From the three *E. coli* PTS systems that are suited for the uptake of glucose, the PTS II^{Glc} and the PTS II^{Man} transport glucose very efficiently (K_m of up to 20µM) while the wildtype PTS II^{Nag} exhibits only a marginal glucose transport capacity that can be dramatically enhanced by a single mutation (Jahreis *et al.*, 2008). While EI and HPr (*ptsI* and *ptsH*, respectively) are common to bacterial phosphotransferase systems, the EII components downstream of HPr are variable in constitution, occurrence and specificity to various substrates.

The glucose-specific PTS of enteric bacteria also links transport to regulation of other relevant proteins: i) At low glucose concentrations, phosphorylated EIIA accumulates and activates membrane-bound adenylate cyclase; the resulting increase in intracellular cAMP levels activates the catabolite activator protein CAP, which in turn causes the transcription of cAMP-CAP dependent genes and operons, a process termed "carbon-catabolite repression" (Park *et al.*, 2006; Bettenbrock *et al.*, 2007). ii) At high glucose concentrations, unphosphorylated EIIA accumulates, which acts as an allosteric inhibitor to proteins involved in the uptake of other carbohydrates such as the lactose permease LacY or the glycerol kinase GlpK, among others. By this, unphosphorylated EIIA prevents the uptake of metabolites which act as inducers for other carbohydrate uptake systems, a process termed "inducer exclusion" (Brückner & Titgemeyer, 2002). For further details about PTS-dependent regulation, see Jahreis *et al.*, 2008. Besides the mentioned PTS systems, glucose can also be internalized by the galactose permease GalP, a proton symporter of the major facilitator superfamily (MFS), and the methyl-galactoside permease MglBAC, belonging to the ATP-binding cassette (ABC) family of transporters (Ferenci *et al.*, 1996). As glucose is not phosphorylated during transport in contrast to PTS-mediated glucose uptake, this is done by the glucose kinase GlkA, an ATP-consuming process that is energetically unfavourable in comparison to the glucose uptake via the PTS (Fig. 10).

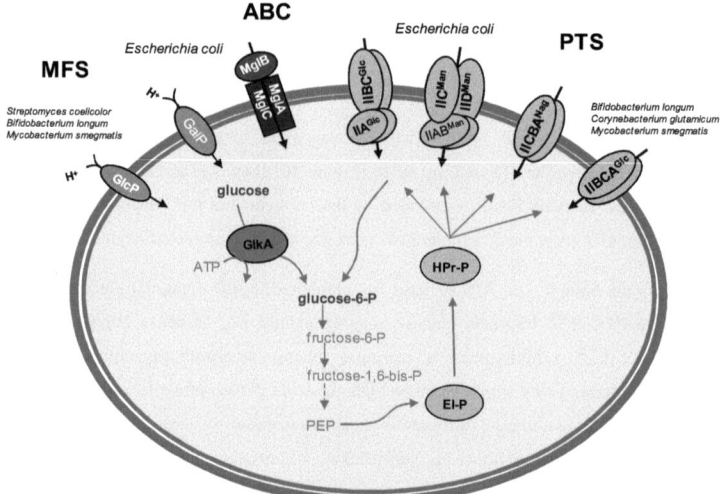

Figure 10: Major glucose uptake systems of *Escherichia coli* K12 and selected Actinobacteria. Note that *S. coelicolor* possesses no glucose-specific PTS II^{Glc} but components of a fructose-PTS II^{Fru} (among others, not shown). The *E. coli* mannose-PTS II^{Man} and *N*-Acetylglucosamine-PTS II^{Nag} are also able to transport glucose. Other ABC (ATP-binding cassette) sugar permeases and transport systems are left out for accessibility. For details, see text and Parche *et al.*, 2007; Titgemeyer *et al.*, 2007; Jahreis *et al.*, 2008. The transporter families are coloured as follows: MFS (major facilitator superfamily), light blue; PTS (phosphotransferase system), yellow; ABC transporter, red. Adapted from Jahreis *et al.*, 2008.

Introduction

For Actinobacteria, research was focused on the GlcP glucose permease of streptomycetes (van Wezel *et al.*, 2005) and the PTS system of corynebacteria (Parche *et al.*, 2001a; 2001b; Moon *et al.*, 2007), while not much was known about sugar uptake systems of mycobacteria, including the pathogenic *M. tuberculosis* and its soil-living relative *M. smegmatis*, or the industrial relevant *B. longum*.

1.5 Proteolysis: An alternative route for carbohydrate and nitrogen sources and its role in pathogenicity

Proteolytic enzymes catalyze the cleavage of peptide bonds and represent approximately two percent of the total number of proteins in all types of organisms (Rao *et al.*, 1998). This means that even in bacteria with strong reduction of genome size, such as in *Mycoplasma* species, several proteases can be found. The multiplicity of these enzymes in cells is caused by the fact that proteases differ significantly in respect to their physiological function. For example, proteases can catalyze the total degradation of their substrate. Resulting amino acids or small peptides can subsequently be used as building blocks for new proteins or as carbon and energy source. In this case, typically a set of rather unspecific endo- and exoproteases work together to ensure a fast and complete hydrolysis of their targets. Proteins used as growth substrates cannot be transported into a bacterial cell. Therefore, proteases are often transported across the cytoplasmic membrane and either bound to the surface of the cell or released into the surrounding medium. The resulting hydrolysis products are subsequently transported into the cell by specific amino acid or peptide uptake systems. Other proteases have a housekeeping function and degrade misfolded proteins or proteins damaged by heat, radiation or other detrimental factors (Wickner *et al.*, 1999; Hengge and Bukau, 2003). Accessory subunits help to identify target proteins and to prevent hydrolysis of functional proteins in this case. Highly specific proteases are also involved in the maturation of preproteins (van Roosmalen *et al.*, 2004) and in the regulatory proteolysis of signal transduction proteins or transcriptional regulators (Gottesman, 1999). Furthermore, in pathogens, proteases are involved in virulence and can damage host cells and tissues (Travis *et al.*, 1995; Armstrong, 2006; Ribeiro-Guimaraes *et al.*, 2007). Due to the multiple functions of proteases, all bacteria have a broad spectrum of these enzymes (for a topical list of peptidases in various species see MEROPS database (Rawlings *et al.*, 2006)). Interestingly, the proteolytic capacity differs significantly between different bacteria.

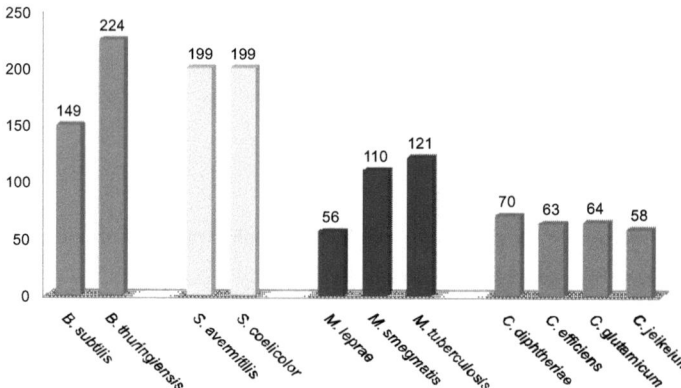

Figure 11: The proteolytic potential of selected Gram-positive bacteria. Shown are the total numbers of predicted putative peptidases according to the MEROPS database (Rawlings et al., 2006). Adapted from Amon et al., 2007.

When major groups of Gram-positive bacteria are compared, namely bacilli, streptomycetes, mycobacteria, and corynebacteria (Fig. 11), it becomes obvious that the spore-forming *Bacillus* species are proteolytically the most active group, especially when the number of proteases is correlated with genome size. Forty eight protease-encoding genes were annotated per Mbp of *Bacillus thuringiensis* chromosomal DNA and 36 genes per Mbp in the genome of *B. subtilis*. The high number of proteins with putative proteolytic functions in *B. thuringiensis* might be the result of its pathogenic lifestyle. This bacterium kills insect larvae by a toxin, which destroys the gut epithelium of the larvae, and uses the decaying bodies as nutrient source (Crickmore, 2005; Bravo *et al.*, 2007). Due to their high proteolysis activity and ability to secrete proteolytic enzymes into the surrounding medium, species like *B. subtilis*, *Bacillus licheniformis* or *Bacillus halodurans* are important industrial producers of alkaliphilic proteases as additives for laundry detergents, amongst various other secreted enzymes (Schallmey *et al.*, 2004; Westers *et al.*, 2004). Despite their large genome sizes of about 9 Mbps, streptomycetes possess only a moderate number of proteases. In the *S. coelicolor* genome approximately 23 protease-encoding genes were found per Mbp chromosomal DNA, in case of *Streptomyces avermitilis* 22 genes. Taking the complex life cycle of streptomycetes and their various metabolic capabilities into account this is surprising. Polypeptides seem to be a less preferred growth substrate of these saprophytic soil bacteria, as streptomycetes can use various other growth substrates including the abundant polysaccharides chitin, xylan and cellulose (Schrempf, 2001; Bertram *et al.*, 2004). Mycobacteria are also only moderately proteolytically active, especially *M. smegmatis*, which features a comparatively small amount of proteins with peptidase function in respect to its genome size of about 7 Mbp. In this organism, only approximately 19 protease-encoding genes were annotated per Mbp of chromosomal DNA. An even more reduced number of proteases was found in *M. leprae* with only 17 proteases encoded per Mbp. The reason might be the lifestyle of this obligate intracellular parasite resulting in severe gene decay termed *reductive evolution* (Eiglmeier *et al.*, 2001). Nevertheless, some of the genes coding for proteins with putative proteolytic activity contribute to the process of human infection and leprosy skin lesions (Ribeiro-Guimaraes *et al.*, 2007). This is also the case for the major human pathogen of the same genus, *M. tuberculosis*, where proteases are studied as important pharmaceutical research targets in concerning infection and virulence (Ribeiro-Guimaraes & Pessolani, 2007). *M. tuberculosis* has the highest number of genes encoding proteases in this genus and approximately 30 of these genes are encoded per Mbp of chromosomal DNA, indicating a more important role of this enzyme class for survival of this bacterium.

For the genus *Corynebacterium*, it was already shown that that at least three different proteases influence the degradation of the PII-protein GlnK, namely FtsH, the ClpCP and the ClpXP protease complex in *C. glutamicum* (Strösser *et al.*, 2004); furthermore, the transcriptional regulator ClgR was reported to control the transcription of ClpCP and other genes involved in proteolysis (Engels *et al.*, 2005). Nevertheless, a detailed comparative genomic analysis of the corynebacterial proteolytic potential with the reported proteome maps of the species *C. jeikeium* (Hansmeier *et al.*, 2007), *C. efficiens* (Hansmeier *et al.*, 2006a) and its putative role in virulence and pathogenicity of *C. diphtheriae* (Hansmeier *et al.*, 2006b) was missing.

1.6 Aim

The aim of this work was to uncover new metabolic pathways and regulatory networks of Gram-positive low G+C species belonging to the phylum Actinobacteria. This approach was based on *in silico* analyses of whole-genome sequences of various industrial relevant and pathogenic model species such as *Corynebacterium glutamicum*, *Bifidobacterium longum*, *Streptomyces avermitilis*, *Mycobacterium smegmatis*, *Mycobacterium tuberculosis* and *Mycobacterium leprae*.

With the help of available genome sequences and homology searches with already described systems of close relatives, the up to now mostly unknown repertoire of nitrogen metabolism-related genes in the genus *Mycobacterium* was to be discovered and putative pathways and regulatory networks to be constructed and compared. Focussing on the non-pathogenic *M. smegmatis*, the putative regulators of nitrogen metabolism GlnR and AmtR were to be characterized by a combined *in silico*, *in vitro* and *in vivo* approach. By this, direct target genes for both regulators were to be verified to construct a regulon for both transcription factor. Furthermore, the interaction partner of GlnR was to be identified in order to gain a more detailed view on nitrogen sensing in Actinobacteria and the role of AmtR in nitrogen-dependent regulation outside of the genus *Corynebacterium* was to be investigated for *M. smegmatis* and *S. avermitilis*.

In respect to carbon metabolism, the up to now mostly unknown repertoire of carbohydrate uptake systems in the genus *Mycobacterium* was to be discovered in order to provide predictions about the comparative growth potential of *M. smegmatis* and *M. tuberculosis* on various carbon sources. Furthermore, the putative glucose permease GlcP and the glucose kinase GlkA of *M. smegmatis* were to be studied in greater detail based on comprehensive sequence and homology analyses. In a similar approach, the genome of the industrially relevant *Bifidobacterium longum* was to be examined. By this approach, a detailed analysis of carbohydrate uptake systems was to result in an overview about the growth potential of this organism on various carbohydrates.

In a comparative genomics *in silico* project, the proteolytic potential of various *Corynebacterium* species based on all available genome sequences was to be unraveled in order to gain a more complete view on the housekeeping protease systems as well as on proteases involved in pathogenicity and virulence of *C. diphtheriae*.

2 Results & Discussion

2.1 Regulation of nitrogen metabolism in *Mycobacterium smegmatis*

2.1.1 Nitrogen-dependent expression of ammonium transport and assimilation proteins depends on the OmpR-type regulator GlnR

Previous work on nitrogen metabolism and its regulation in mycobacteria concentrated strongly on glutamine synthetase in *M. tuberculosis*, which is essential in this bacterium and consequently an important drug target (Tullius *et al.*, 2003). Additionally, studies of the *glnE* gene product adenylyltransferase, which is crucial for posttranslational modification and regulation of glutamine synthetase, and the *glnD*-encoded uridylyltransferase, which is putatively involved in nitrogen signal transduction, were carried out (Harth *et al.*, 1994; 2005; Parish & Stoker, 2000; Read *et al.*, 2007). In *M. smegmatis*, GlnR binding sites were identified upstream of the *glnA*, *amtB* and *amt1* genes. As in other actinomycetes (Jakoby *et al.*, 2000), the *M. smegmatis amtB* gene forms an operon with downstream *glnK* and *glnD* genes. A core ammonium assimilation regulon is controlled by *M. smegmatis* GlnR and regulates transcription of *glnA*, which encodes glutamine synthetase, the ammonium transporter-encoding genes *amtB* and *amt1* and corresponding signal transduction components encoded by *glnK* and *glnD* (Fig. 12).

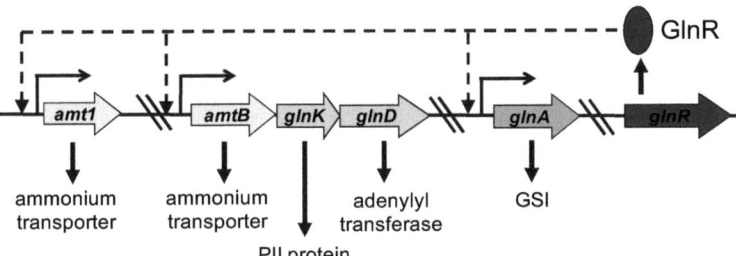

Figure 12: The *M. smegmatis* GlnR regulon. For details about gene functions, see text. Adapted from Amon *et al.*, 2008a.

The signal transduction to *M. smegmatis* GlnR is unclear, a situation which is similar to that in *S. coelicolor*. Based on the conserved phosphorylation domain and analogous to the situation in *E. coli* (Yoshida *et al.*, 2006), involvement of a protein kinase is assumed. However, the corresponding protein is unknown. Based on the data obtained for *M. smegmatis* and previous studies focusing on *S. coelicolor* (Fink *et al.*, 2002; Tiffert *et al.*,

Results & Discussion

2008; Wray & Fisher, 1993; Wray *et al.*, 1991) and *Amycolatopsis mediterranei* (Yu *et al.*, 2006; 2007), GlnR seems to be a major regulator of ammonium assimilation in actinomycetes (see also Fig. 16). This conclusion is supported by the results of electrophoretic mobility shift assays, which showed that *S. coelicolor* GlnR is able to bind to the *glnA* promoter regions of *B. longum*, *Frankia* sp. strain EAN1, *M. tuberculosis*, *Nocardia farcinica*, *Nocardioides* sp. strain JS614, *Propionibacterium acnes* and *Rhodococcus* sp. strain RHA1 *in vitro* (Tiffert *et al.*, 2008).

In contrast to the situation in most other actinomycetes, GlnR does not play a role in nitrogen control in the genus *Corynebacterium*. In this genus, no GlnR homologs were observed; instead, AmtR is the central nitrogen control protein (Walter *et al.*, 2007). Compared to the recently described extended *S. coelicolor* GlnR regulon (Tiffert *et al.*, 2008), *M. smegmatis* GlnR has a reduced number of target genes. For the NADP-dependent glutamate dehydrogenase gene *gdhA* (*msmeg_5442*), the urease operon (*msmeg_2627* to *msmeg_3622*), and the nitrite reductase genes (*msmeg_0427* and *msmeg_0428*), genes that are under the control of GlnR in *S. coelicolor*, no GlnR *cis* elements were found in this study.

2.1.2 The role and function of AmtR in *M. smegmatis* and *S. avermitilis*

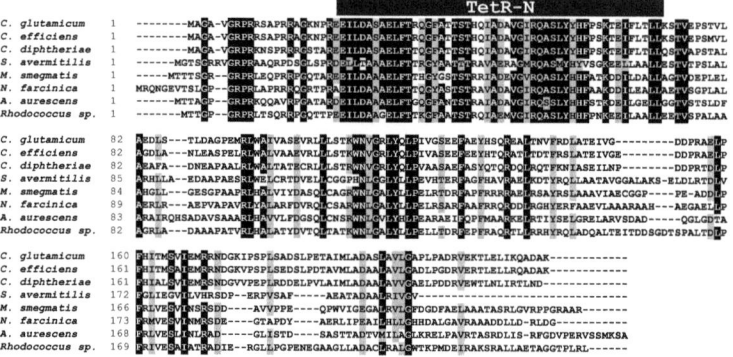

Figure 13: Sequence alignment of selected AmtR-homologs. Amino acid residues identical in all sequences are shaded in black, other conserved amino acids in gray. Black bar marks the conserved N-terminal TetR DNA binding domain. Adapted from Muhl *et al.*, 2009. For full species names see appendix.

Preliminary data indicate that there may be nitrogen-dependent regulation of the nitrite reductase and urease, and AmtR might be an interesting candidate for a second nitrogen regulator, considering the presence of a gene encoding an AmtR homolog in the genome of *M. smegmatis* and the fact that for *C. glutamicum* the urease-encoding genes - among many others - are under the control of the AmtR repressor (Beckers *et al.*, 2005). Together with *N. farcinica*, *Rhodococcus* sp. strain RHA1 and *Arthrobacter aurescens*, *M. smegmatis* seems to be one of the few actinomycetes besides the members of the genus *Corynebacterium* to have an AmtR homolog (Figure 13).

Co-occurrence searches done with the STRING server at the EMBL, Heidelberg, suggested a putative role of AmtR in repression of an operon that might be involved in amino acid uptake and assimilation, which is conserved in the genomes of *M. smegmatis* and close relatives. This operon comprises, among two hypopthetical reading frames of unknown function, genes encoding a predicted amino acid permease and enzymes putatively involved in amino acid degradation (Table 1).

Table 1: Distribution of AmtR and putative target genes in *M. smegmatis* and close relatives. Hypothetical proteins msmeg_2185 & msmeg_2186 not shown. For *C. efficiens*, AmtR binding motifs in the upstream regions are predicted. For full species names, see appendix.

	amino acid permease	urea hydrolase	amidase family protein	AmtR
M. smegmatis	msmeg_2184	msmeg_2187	msmeg_2189	msmeg_4300
C. efficiens*	ce0711	ce0713	ce0710	ce0939
S. avermitilis	sav6709	sav6698	sav6697	sav6701
Rhodococcus sp.	RHA1_ro06919	RHA1_ro06922	RHA1_ro02136	RHA1_ro06918
N. farcinica	nfa22220	nfa22190	nfa22180	nfa22230
A. aurescens	AAur_0190	AAur_0187	AAur_0186	AAur0192
C. glutamicum	n/a	n/a	n/a	cg0986
C. diphtheriae	n/a	n/a	n/a	DIP0846

Early experimental results not only confirm the bioninformatics predictions but also point to the hypothesis that regulation of this operon might in fact be a result of dual regulation, namely by AmtR and the main nitrogen regulator GlnR (Fig. 14).

Results & Discussion

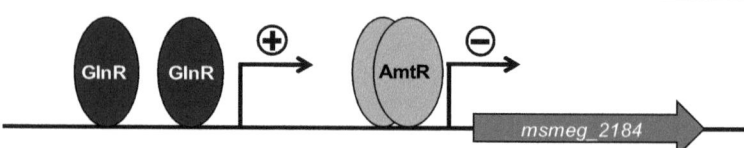

Figure 14: Model of dual transcriptional regulation by GlnR and AmtR. Note that AmtR as a member of the TetR family is a transcriptional repressor that is supposed to bind to DNA as a dimer while GlnR, an activator, binds to DNA in a mode termed "galloping model" (Yoshida *et al.*, 2006). Binding sites might actually overlap.

AmtR and GlnR double mutants support this hypothesis for both *M. smegmatis* and *S. avermitilis* (Dr. Yinhua Lu, personal communication). Ongoing transcriptome studies and footprinting of the exact DNA binding sites for AmtR and GlnR will result in more evidence and an expanded model of the dual regulation of nitrogen uptake and metabolism in *M. smegmatis* and *S. avermitilis*.

2.1.3 The search for the GlnR interaction partner

Suggesting OmpR as a model system for GlnR, it is to be expected that GlnR as a typical response regulator of a bacterial two-component system is being phosphorylated. Indeed, in an alignment with several homologous GlnR proteins, various conserved aspartate residues have been identified that putatively get phosphorylated by a yet unknown histidine kinase (Figure 15).

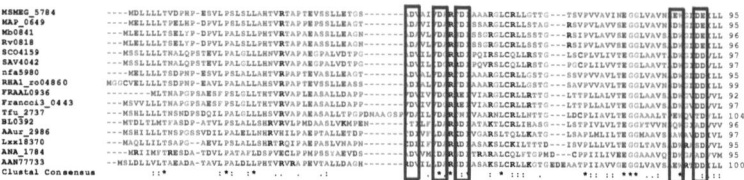

Figure 15: Alignment of N-termini of selected GlnR homologs. Given are the exact annotation numbers according to the full genome sequences of the respective organisms. Red, conserved aspartate residues.

Up until now, no specific phosphorylation site could be determined, as an exchange to alanine of each of the three C-terminal aspartate residues results in a complete loss of function of the regulator, while phosphorylation assays in fact point to a phosphorylation of the protein (N. Jessberger, personal communication). In parallel, a genomic screening was undertaken based on co-occurrence searches and following prerequisites: a) The putative histidine kinase is found in every actinobacterial genome that also features a GlnR homolog; b) the kinase must not be found in corynebacterial genomes as AmtR is the regulator of nitrogen metabolism in this genus; and c) it is supposed to be an orphan kinase to the orphan regulator GlnR, i.e. the kinase may not occur as a typical two-component system pair with its cognate response regulator genomically. The screening resulted in a comprehensive analysis of histidine kinases encoded in the genomes of *M. smegmatis* and other Actinobacteria, which is presented in table 2.

Table 2: Analysis of the histidine kinases of *M. smegmatis*. Given are the predicted open reading frames with their corresponding annotation number for the respective genome and their identity in percentage according to BLASTP results. On the right, the putative role based on literature research and homology searches is given. n/a, not available.

M. smegmatis	other mycobacteria	*N. farcinica*	streptomycetes	corynebacteria	putative role
MSMEG_5158	n/a	NFA17980 (69%)	SCO1217 (49%) SAV7118 (50%)	n/a	LytS (regulation of cell autolysis)
MSMEG_2793 MSMEG_5663 MSMEG_0246	yes (all)	NFA6630 (46%)	n/a	n/a	PrrB (mult. copies)
MSMEG_5304	n/a	NFA46340 (54%)	SCO5435 (48%) SAV2816 (47%)	CG0089 (33%)	CitA regulating citrate/malate metabolism?
MSMEG_3239 (no orphan)	MAP_3274 (65%)	NFA54920 (48%)	SCO2121 (38%) SAV6081 (39%)	n/a	MSMEG_3240 LuxR-type regulator, conserved across species...
MSMEG_5870 MSMEG_4989	yes (all except ML)	NFA5450 (54%)	SAV4197 (46%) SCO4021 (49%)	CG2887 (44%) [JK0342 (48%), DIP1935 (45%), CE2493 (41%)]	PhoR ribonucleotide biosynthesis (mult. copies)
MSMEG_0106 (no orphan)	yes (all except ML)	n/a	SCO1369 (33%)	CG2201 28%	?
MSMEG_6864	n/a	NFA12320 (32%)	SCO7562 (35%)	CG3388 25%	?
MSMEG_2248	n/a	n/a	n/a	n/a	?
MSMEG_4307	yes (all)	NFA16340 (64%)	n/a	CG2457 (49%) DIP1680 (53%), JK0667 (51%), CE2135 (50%)	?
MSMEG_4211	n/a	n/a	SCO1137 (43%)	n/a	similar MSMEG_530

Results & Discussion

no orphan, cit?					4
MSMEG_1918	**yes (all)**	**NFA45810 (64%)**	**SCO5239 (46%) SAV3017 (46%)**	**n/a**	**? (VERY good candidate)**
MSMEG_2915 (no orphan!)	yes (all except ML)	n/a	…	…	similar to MSMEG_4989 /5870…
MSMEG_0980 (no orphan)	n/a	n/a	…	…	similar to MSMEG_2248
MSMEG_3448 (no orphan)					
MSMEG_2804 (no orphan)					
MSMEG_5241	**yes (all except ML)**	**NFA28940 (63% BLASTN)**	**SCO0203 (61%) SAV4257 (59%)**	**n/a**	**"GAF family protein"; orphan, also very good candidate**
msmeg_0854	n/a	nfa7820	sco6163/6424/5784	n/a	
msmeg_1493 (no orphan)	n/a	n/a	n/a	n/a	
msmeg_4968 (no orphan)	n/a	n/a	sco4597/4598	n/a	
msmeg_0936 no orphan!	rv0490 (all)	nfa51870	sco4229/sav3973	cg0483/ce0424	"SenX3"

From all identified histidine kinases, two candidates were selected for experimental validation, namely Msmeg_1918 and Msmeg_5241. Experimental work is ongoing, but deletion mutants of both kinases showed no specific influence on nitrogen-dependent regulation of GlnR target genes (personal communication, N. Jessberger).

2.2 Comparative genomic analysis of nitrogen metabolism and control in mycobacteria

M. smegmatis is equipped with a variety of genes enabling the uptake and assimilation of nitrogen sources. Compared to the fast-growing *M. smegmatis*, all slow-growing pathogenic members of the genus exhibit a reduced number of genes encoding proteins for nitrogen uptake and assimilation. This is also due to the fact that *M. smegmatis* seems to have acquired an astonishingly wide range of nitrogen-related genes and gene regions via horizontal gene transfer from a variety of other bacteria, such as *Agrobacterium, Burkholderia,* and *Pseudomonas* species. This includes among others a second urease operon, additional ammonium transporters, and a broad variety of glutamine synthetases of various classes and origins. According to the genomic data, *M. smegmatis* is capable of the active uptake and assimilation of a comparatively wide range of substrates for the extraction of ammonium and further assimilation into central metabolites such as glutamate and glutamine (Amon *et al.*, 2009), which is in good concordance to the situation found for the uptake and assimilation of carbohydrates in *M. smegmatis* (Titgemeyer *et al.*, 2007) and thus exhibits a similar repertoire of nitrogen-related genes to that of *C. glutamicum* (Burkovski, 2007; Hänssler & Burkovski, 2008).

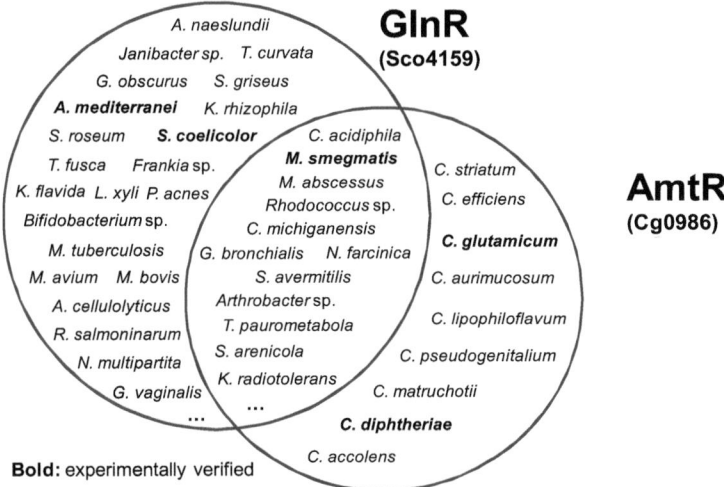

Figure 16: Distribution of putative nitrogen-dependent transcription regulators GlnR and AmtR in selected Actinobacteria, based on current genomics data. Bold species names indicate experimental evidence for nitrogen control by the corresponding protein. In *M. smegmatis* only the function of GlnR is characterized, while AmtR function remains unclear. The references for experimentally verified systems are as follows: Amon *et al.*, 2008a; Fink *et al.*, 2002; Jakoby *et al.*, 2000; Nolden *et al.*, 2002; Yu *et al.*, 2006. Adapted from Amon *et al.*, 2009. For a full list of species names, see appendix.

Another interesting fact is the co-occurrence of homologs of both regulators of the nitrogen metabolism in actinomycetes in the genome of *M. smegmatis*, namely AmtR and GlnR (Amon *et al.*, 2009). While the transcriptional repressor AmtR is the global regulator of nitrogen metabolism in corynebacteria (Walter *et al.*, 2007), this function was only recently shown for the *M. smegmatis* GlnR and its respective target genes *amt1*, *amtB*, and *glnA* (Amon *et al.*, 2008a); as only homologs of GlnR are found in the genomes of other mycobacteria (Fig. 16), the role of AmtR for *M. smegmatis* remains to be further explored.

All mycobacteria investigated exhibit the subunits for a respiratory nitrate reductase (which are nonfunctional pseudogenes in *M. leprae*), while especially the tuberculoid members possess multiple homologs of the *E. coli* nitrite/nitrate antiporters, NarK and NarU (Amon *et al.*, 2009). For *M. tuberculosis* it has already been shown that nitrate respiration plays an important role during hypoxia (Sohaskey, 2008), but the additional occurrence of a nitrite reductase, besides its role in detoxification by reduction of nitrite, points to a possible involvement of the enzyme in the complete reduction of nitrate to ammonium and following assimilation, which has been demonstrated for *M. smegmatis* (Khan *et al.*, 2008).

M. leprae unsurprisingly reveals the strongest reduction of nitrogen metabolism-related genes as a process of gene decay termed '*reductive evolution*' (Gómez-Valero *et al.*, 2007), resulting in a minimal set of genes required for a functional nitrogen metabolism. This set comprises the genes of the GS/GOGAT pathway (*gltBD*, *glnA*) as well as the GS ATase (*glnE*) (Amon *et al.*, 2009). While no nitrogen-specific transport systems besides one nitrite/nitrate antiporter (NarK1) have been found, the genome of *M. leprae* features putative amino acid and oligopeptide transporters and permeases; these substrates could very well represent the main nitrogen sources, taking into account the lifestyle of *M. leprae* as an obligate intracellular parasite (Sassetti *et al.*, 2003; Vissa & Brennan, 2001).

2.3 The glucose permease and glucose kinase of *M. smegmatis*

The protein database of the *M. smegmatis* genome was scanned with several known glucose permease protein sequences using BLASTP as implemented at The Institute for Genomic Research (TIGR) (Parche *et al.*, 2006; van Wezel *et al.*, 2005). Only one gene *(msmeg4187)* was found that encodes a predicted glucose permease of the major facilitator super family (MFS) (Pao *et al.*, 1998). The deduced protein shares 53% amino acid identity to the well-characterized glucose permease GlcP of Streptomyces coelicolor (Titgemeyer *et al.*, 2007; van Wezel *et al.*, 2005). Analysis of the gene locus of *msmeg4187* (hereafter named *glcP*) indicated that the gene is transcribed as a monocistronic mRNA (Pimentel-Schmitt *et al.*, 2009). *glcP* is flanked downstream by a putative dicistronic *hisEG* operon of genes involved in histidine biosynthesis, and upstream by an unknown open reading frame. While the *hisEG* operon is conserved in mycobacteria, it appears that the slow-growing mycobacteria, e.g. *M. tuberculosis*, *M. bovis*, and *M. avium*, do not have an obvious homologue of *glcP*. The multiple alignment includes four homologous glucose permeases which have been experimentally described (Parche *et al.*, 2006; van Wezel *et al.*, 2005; Weisser *et al.*, 1995; Zhang *et al.*, 1989). The range of protein identity is from 34 to 53%, showing 67 conserved residues in these sequences. Hydrophobicity profiles showed that all five sequences have 12 predicted transmembrane helices that are at equivalent positions (Pimentel-Schmitt *et al.*, 2009). To establish the phylogenetic relationships of GlcP, 19 protein sequences of the MFS were selected which have been characterized by experimental investigation including those from diverse mycobacteria (Fig. 17).

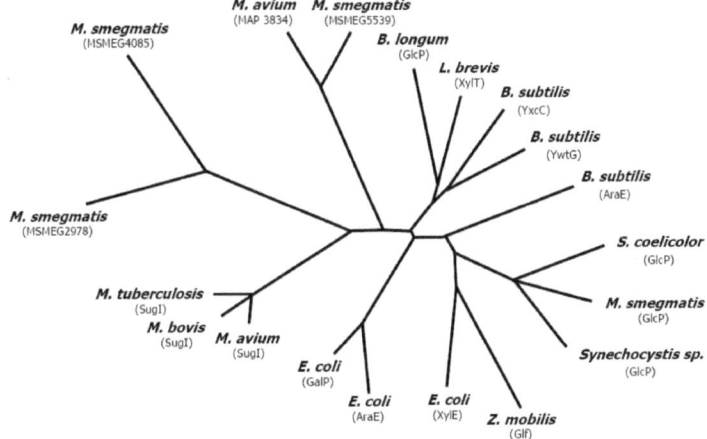

Figure 17: Phylogenetic tree of the *glcP* family. An unrooted phylogenetic tree was computed with the CLUSTALW software making use of the implemented neighbor joining method with the function for evolutionary distance correction. Evolutionary distances are proportional to the branch length. 19 protein sequences were selected as indicated in the figure. The references for experimentally verified transporters are as follows: Chaillou *et al.*, 1998; Krispin *et al.*, 1998; Martin *et al.*, 1994; Parche *et al.*, 2006; van Wezel *et al.*, 2005; Weisser *et al.*, 1995; Zhang *et al.*, 1989. Adapted from Pimentel-Schmitt *et al.*, 2009. For full species names see appendix.

GlcP clusters with the *S. coelicolor* homolog and the one from the cyanobacterium *Synechocystis* (van Wezel *et al.*, 2005; Zhang *et al.*, 1989). Interestingly, the xylose permease XylE of *E. coli* is more closely related to GlcP than expected, while the bifidobacterial GlcP is more distant but closely associated with another xylose permease, XylT of *Lactobacillus*

Results & Discussion

brevis (Chaillou *et al.*, 1998). This suggests that glucose and xylose transporter genes may have evolved following two different branches from the common origins. The glucose-specific transport protein GlcP from *M. smegmatis* encoded by open reading frame *msmeg4187* is the first reported mycobacterial sugar permease. Interestingly, no GlcP homolog is present in slow-growing mycobacteria, notably *M. tuberculosis* and *M. leprae* (Titgemeyer *et al.*, 2007). In these bacteria, glucose may enter either by simple diffusion, which could be efficient enough to meet the nutritional demands of these extremely slow-growing bacteria (generation time of weeks), or through another, yet unknown permease (Titgemeyer *et al.*, 2007).

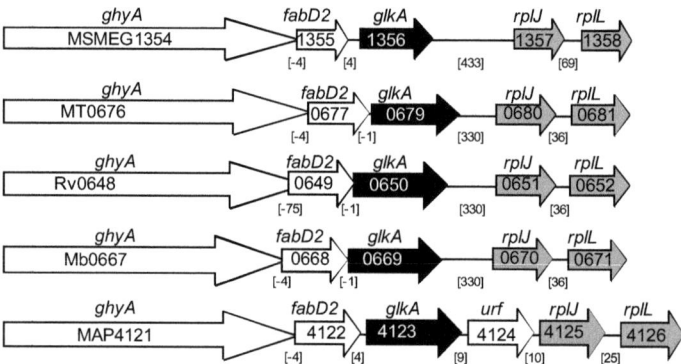

Figure 18: Genetic organization of the *glkA* gene in *M. smegmatis*. The highly conserved depicted *glkA* region (black) includes the genes *ghyA* (encoding a glycosyl hydrolase), *fabD2* (malonyl-CoA transacylase), and the genes for the ribosomal proteins L10 and L7/L12 (gray). Note that *M. avium paratuberculosis* additionally carries a gene encoding a protein of unknown function (*urf*, MAP4124). The genomes of *M. smegmatis* mc^2 155 (MSMEG), *M. tuberculosis* H37Rv (Rv), *M. tuberculosis* CDC1551 (MT), *M. bovis* subsp. *bovis* AF2122/97 (Mb), and *M. avium paratuberculosis* (MAP) were sequenced and annotated by TIGR. The arrows indicate length and transcriptional orientation of annotated genes and open reading frames. Numbers in brackets show the lengths of intergenic regions in bp. Adapted from Pimentel-Schmitt *et al.*, 2007.

Furthermore, genomes of *M. smegmatis* and *M. tuberculosis* H37RV were screened for the presence of putative glucose kinase genes. A protein BLASTP search with the Glk protein sequence of *S. coelicolor* revealed the presence of seven homologues in the genome of *M. smegmatis* (Pimentel-Schmitt *et al.*, 2007). These sequences are members of the ROK family, which comprises bacterial transcription factors and sugar kinases (Titgemeyer *et al.*, 1994a; 1994b). Three of the seven genes are sugar kinases. Of these, the gene product MSMEG1356 exhibited with 37% identical and 58% similar amino acid residues the closest homologue. Analyzing the genome of *M. tuberculosis*, three ROK family members have been found, of which one gene, Rv0650, was close to *glk* of *S. coelicolor* and MSMEG1356. Hence, MSMEG1356 (hereafter termed *glkA*) was chosen as the prime candidate for a possible glucose kinase (Pimentel-Schmitt *et al.*, 2007). A comparison of the *glkA* region with other mycobacterial genomes revealed a very similar genetic context (Fig. 18), in which *glkA* is the third gene of a putative tricistronic operon.

The first one is a possible glycosyl hydrolase, suggesting that the enzyme together with GlkA participates in the same pathway, which could be the metabolism of glucose-containing saccharides. The second gene appears to encode a putative malonyl-CoA transacylase found as a unique conserved mycobacterial gene (Huang *et al.*, 2006). Two genes that encode the ribosomal proteins L10 and L7/L12 are located immediately downstream of *glkA*. Such a conserved gene order is a good hint for equivalent functions in closely related species. A multiple alignment of homologous glucose kinases from diverse bacteria reveals that GlkA shows 71% protein identity to the one from the *M. tuberculosis* H37Rv reference strain and 28–52% to the others that were chosen from related actinomycetes and from *Bacillus subtilis* (Pimentel-Schmitt *et al.*, 2007). An unrooted tree shows that GlkA is most closely related to the mycobacterial homologues, then to SCO1077 (GlkII) of *S. coelicolor* and then equally distant to other glucose kinases (Pimentel-Schmitt *et al.*, 2007).

The occurrence of a glucose kinase gene in all so far sequenced mycobacterial genomes indicates an important function. Multiple alignment analysis revealed 39 amino acid residues, which are conserved in all six selected species. Among these residues are several that have been identified to participate in ATP, glucose, and zinc binding in distantly related species (Hansen *et al.*, 2002; Lunin *et al.*, 2004; Mesak *et al.*, 2004; Schiefner *et al.*, 2005; Titgemeyer *et al.*, 1994a). Future research should be focused on the question: which protein or metabolic pathways in *M. smegmatis* are affected by *glkA*? In particular, it will be important to examine whether *glkA* plays a direct metabolic role in glucose utilization and/or carbon regulation. So far, GlkA may not be required to maintain a functional virulence cycle. This at least can be inferred from a global microarray-based analysis of *M. tuberculosis* (Sassetti *et al.*, 2003; Sassetti & Rubin, 2003). Assuming that specific inhibitors of protein kinases have been successfully developed for therapeutic usage against a variety of diseases (Shawver *et al.*, 2002), future investigation for drug development could be directed to the selection of GlkA of mycobacterium, as a means of interfering with growth and possible regulatory processes. *M. smegmatis* and also the slowgrowing mycobacterial species possess a glucose kinase that converts the incoming glucose to glucose-6-phosphate (Pimentel-Schmitt *et al.*,

Results & Discussion

2007; Titgemeyer et al., 2007). Apart from its catalytic function, this enzyme confers global carbon regulation in closely related streptomycetes (Angell et al., 1994). It has been shown in *S. coelicolor* that glucose kinase binds to glucose permease GlcP (van Wezel et al., 2007). This protein-protein interaction is thus thought to influence the regulatory function of glucose kinase. For this reason, it would be worthwhile to study, whether GlcP and GlkA of *M. smegmatis* also interact with each other and whether this has similar consequences regarding nutritional control. This in turn would help to understand how nutrition and virulence may be linked in mycobacteria.

2.4 Carbohydrate transport systems of mycobacteria

It has been widely documented that *M. smegmatis* can grow on many carbon sources such as polyols, pentoses and hexoses (Edson, 1951; Franke & Schillinger, 1944; Izumori et al., 1978). Multiple inner membrane transport systems for all three of these classes of carbohydrates by bioinformatic analysis have been identified (Titgemeyer et al., 2007). This provides the molecular basis for the adaptability of *M. smegmatis* to different environments in the soil and water. Often, the integrated bioinformatic approach enabled us to propose a specific substrate for particular uptake proteins (Fig. 19). Since the specificity of transport proteins can be altered by the modification of a few residues, the suggested substrates rather represent a hypothesis for experiments such as transport measurements with gene deletion mutants or analysis of the induction of gene expression. Analysis of the induction of both the transport activity and transcription of genes confirmed the substrate predictions for a fructose- and glucose-specific PTS as well as for the predicted glycerol operon. Since glucose uptake was constitutive, at least one more system for glucose transport must occur in *M. smegmatis*. This was predicted to be GlcP. Indeed, cloning and heterologous expression of *glcP* in *E. coli* revealed that it is a glucose-specific permease (Pimentel-Schmidt et al., 2009).

The discovery of homologs of all components of a PTS in *M. smegmatis* contradicts a previous report that did not find biochemical evidence for the existence of a PTS (Romano et al., 1970) and many repeating statements (Connell & Nikaido, 1994; Content et al., 2005; Ratledge & Stanford, 1982). Components of the PTS play a key role in the global control of sugar metabolism to achieve the hierarchical utilization of carbon sources in bacteria (Brückner & Titgemeyer, 2002), where two different mechanisms have evolved. In *E. coli* and other closely related gram-negative bacteria, the enzyme IIAGlc is dephosphorylated mainly under repressing conditions and mediates inducer exclusion. Under nonrepressing conditions, phosphorylated IIAGlc stimulates cyclic AMP (cAMP) synthesis and thereby triggers the activation of catabolite- repressed genes by a global regulator, the cAMP-dependent catabolite activator protein (see 1.4).

In low G+C Gram-positive bacteria such as *Bacillus subtilis*, HPr is a central switch of carbon catabolite repression. Under repressing conditions, HPr is phosphorylated mainly at serine 46 by a unique HPr kinase/phosphatase mediating inducer exclusion and carbon catabolite repression/activation (Brückner & Titgemeyer, 2002; Monedero *et al.*, 2007; Reizer *et al.*, 1998).

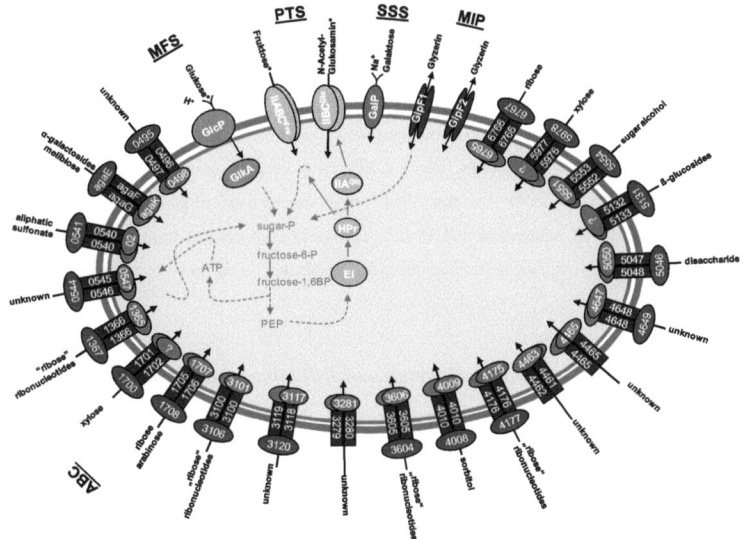

Figure 19: Predicted sugar transport systems of *M. smegmatis*. Shown are the permeases of the ABC (orange), PTS (yellow), MIP (violet), MFS (green) and SSS (blue) families. The derived putative substrates are inferred from *in silico* analyses in combination with experimental data. Adapted from Titgemeyer *et al.*, 2007.

Under non-repressing conditions, HPr is phosphorylated at histidine 15 and activates PTS-dependent sugar transport, glycerol kinase, and substrate-specific regulators. The apparent absence of a protein in *M. smegmatis* similar to the HPr kinase/phosphatase argues against the mechanism found in low-GC Gram-positive bacteria. On the other hand, *M. smegmatis* apparently does not produce proteins with significant sequence similarities to the cAMP receptor protein (CRP) (catabolite activator protein) of *E. coli*, which is crucial for

Results & Discussion

carbon catabolite repression in Gram-negative bacteria. By contrast, the coordination of the few operons involved in the uptake and degradation of carbohydrates by *M. tuberculosis* (Titgemeyer *et al.*, 2007) may not require a global control mechanism, as suggested by the lack of PTS homologs. Alternatively, a completely different mechanism for the global control of carbon metabolism may have evolved in *M. tuberculosis* to adapt to its specific environment inside the phagosome of human macrophages. Indeed, M. tuberculosis has eight orthologs of CRP-like transcriptional regulators (McCue *et al.*, 2000), one of which, Rv3676, was experimentally described (Bai *et al.*, 2005). Furthermore, the large number and the different subcellular localization of the 15 putative nucleotide cyclases in *M. tuberculosis* imply that this organism may have the ability to sense and respond to many intracellular and extracellular signals through a second messenger system based on cyclic nucleotide monophosphates (McCue *et al.*, 2000). This is in strong contrast to *E. coli* and other Gram-negative bacteria, which have only one CRP and one adenylate cyclase. CRP homologs have been identified in streptomycetes (Derouaux *et al.*, 2004), where the regulator plays a role in germination, and in corynebacteria, where CRPs have been associated with global carbon regulation (Moon *et al.*, 2007). Although the potential mechanisms of global control of carbon metabolism in both *M. smegmatis* and *M. tuberculosis* are not evident from the bioinformatic analysis of their genomes, these findings provide hypotheses for further experiments.

2.5 Carbohydrate transport systems of *Bifidobacterium longum*

In silico analysis was done by conducting BLASTP analyses at the BLAST server of the TIGR institute following previously described strategies (Bertram *et al.*, 2004). The detected gene products from *B. longum* NCC2705 and those of the adjacent genes were then searched with BLASTP at various genome servers, including the one of the Transporter Classification Database (TCDB; www.tcdb.org), in order to identify all carbohydrate transporter genes (Saier *et al.*, 2006). The search led to the identification of 19 *B. longum* loci predicted to encode carbohydrate transporters (Parche *et al.*, 2007; Fig. 20). They belong to the ATP binding cassette family (ABC), the phosphotransferase system (PTS), the major facilitator superfamily (MFS), the glycoside-pentoside-hexuronide cation symporter family (GPH), and the major intrinsic protein family (MIP). The majority was formed by 13 ABC permeases (ATP-binding cassette), which include porters for di-, tri- and higher order oligosaccharides. Three permeases of the MFS, which include proton symporters and facilitators, are predicted to transport glucose, lactose, and sucrose. Also found was one

Results & Discussion

complete glucose-specific PTS. It comprises the general phosphotransferases enzyme I (EI) and HPr, which operate in global carbon regulation in many bacteria (Brückner & Titgemeyer, 2002), and a single predicted glucose- specific enzyme II permease. Finally, a putative glycerol permease of the MIP and one permease of the GPH family completed the list of detected NCC2705 sugar transport systems (Parche *et al.*, 2007). To gain more hints on possible substrates, the genomic neighborhood was inspected for associated metabolic genes, which are usually colocalized in an operon structure, and for possible regulators, which often occur adjacent to the target operon.

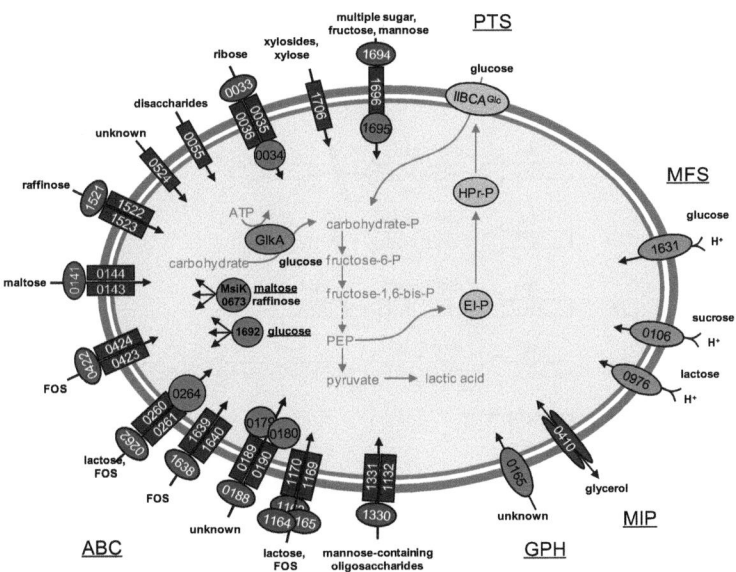

Figure 20: Predicted sugar transport systems of *B. longum* NCC2705. Shown are the permeases of the ABC (red), PTS (yellow), MFS (green), MIP (violet) and GPH family (blue). FOS, fructooligosaccharide. The derived putative substrates are inferred from *in silico* analyses in combination with experimental data. Experimentally verified systems are denoted by an asterisk. Adapted from Parche *et al.*, 2007.

The combined analysis of the carbohydrate transporter systems of *B. longum* NCC2705 revealed 19 transport systems and provided a number of interesting conclusions: (1) *B. longum* can transport a variety of disaccharides and oligosaccharides like oligofructose, which are described as growth-promoting prebiotics. Compared to the initial publication of the genome sequence (Schell *et al.*, 2002), in which 7 oligosaccharide systems were presented, our more extensive analysis revealed that up to 15 such systems may exist in NCC2705. (2) More than half of the transport systems are ATP-dependent ABC-type permeases, a feature that is also found in other actinomycetes such as *S. coelicolor* (Bertram *et al.*, 2004) and *Mycobacterium smegmatis* (Titgemeyer *et al.*, 2007). (3) PTSs are present in bifidobacteria and occur either as glucose- or fructose-specific PTS. It remains to be established whether these PTS components serve additional regulatory function besides sugar transport as they do in other bacteria (Brückner & Titgemeyer, 2002; Rigali *et al.*, 2006).

2.6 Comparative genomic analysis of the proteolytic potential in corynebacteria

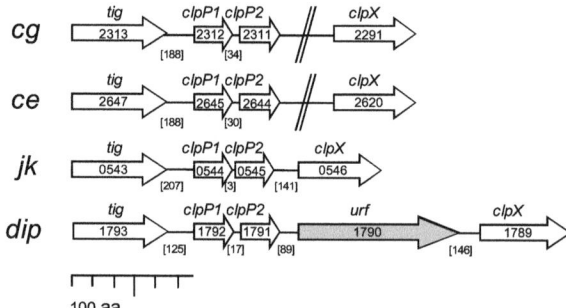

Figure 21: As an example of a conserved proteolytic system, the genetic organization of *clp* gene clusters in corynebacteria is shown. Arrows indicate the lengths and transcriptional orientations of annotated genes and predicted ORFs. Genes are shown by their annotation number, with the prefix of the respective organism. The gene names are assigned according to the annotations given by TIGR (http://www.tigr.org). Numbers in square brackets refer to the intergenic distance between two genes. Gene designations are as follows: *urf*, unknown reading frame; *tig*, trigger factor; *clpP1*, *clpP2*, proteolytic subunits; *clpX*, ATPase subunit; *cg*, *C. glutamicum*; *ce*, *C. efficiens*; *jk*, *C. jeikeium*; *dip*, *C. diphtheriae*. Adapted from Amon *et al.*, 2008b.

Based on published genome sequences of the *Corynebacterium* species *C. diphtheriae* (Cerdeno-Tarraga *et al.*, 2003), *Corynebacterium efficiens* (Fudou *et al.*, 2002), *C. glutamicum* (Ikeda & Nakagawa, 2003; Kalinowski *et al.*, 2003) and *C. jeikeium* (Tauch *et al.*, 2005), a comparative genomics analysis of genes encoding proteins with putative proteolytic functions was performed. A complete overview of protease genes found in each of the corynebacterial genomes is given in Amon *et al.*, 2008b. A total number of 53, 51, 56, and 42 protease-encoding genes was predicted in the genomes of *C. glutamicum*, *C. diphtheriae*,

C. efficiens, and *C. jeikeium.* Correlated to the genome size, *C. efficiens* has the lowest density of protease encoding genes on the chromosome (less than 20 per Mbp), followed by *C. jeikeium* (approximately 23 genes per Mbp), *C. glutamicum* (about 29 genes per Mbp) and *C. diphtheriae* (more than 30 genes per Mbp). A core subset of 39 homologous proteases is common to all corynebacterial species. These findings are in good accordance with the numbers recently published for the closely related mycobacterial genus (Ribeiro-Guimaraes *et al.*, 2007), comprising a similar set of highly conserved homologous proteases.

Compared to other bacteria, corynebacteria seem to have only moderate proteolytic activity. Nevertheless, proteases obviously play a crucial role in many cellular processes. First investigations of *C. glutamicum* Clp protease complexes (Fig. 21) and FtsH hint to interesting regulatory functions of these AAA+ enzymes in this species. However, knowledge on this class of proteases is far from completeness and for example even the degradation signals are still unknown. The same is true for the function of putative degradation tags or amino acid residues influencing protein stability (Mogk *et al.*, 2007). As discussed above, already the increased density of protease-encoding genes in the *C. diphtheriae* genome compared to other corynebacterial genomes and their partial localization on genomic islands suggest a role of these enzymes in pathogenicity. In fact, *C. diphtheriae* sortases were already shown to be crucial for pili assembly and adhesion to hosts cells, an effect also observed for two other secreted *C. diphtheriae* proteases. However, also in this case, only limited information is available. Future approaches might shed more light on the diverse function of known and uncharacterized proteases in different corynebacteria and hopefully elucidate the up to now completely unknown target recognition mechanisms of these enzymes.

3 Summary / Zusammenfassung

3.1 Summary

Based on available genome sequences and homology searches with already described systems of close relatives, the up to now mostly unknown repertoire of nitrogen metabolism-related genes in the genus *Mycobacterium* was discovered and putative pathways and regulatory networks have been constructed and compared. Focussing on the non-pathogenic *Mycobacterium smegmatis*, the regulator of nitrogen metabolism GlnR was characterized by a combined *in silico*, *in vitro* and *in vivo* approach. By this, direct target genes for the regulator were verified and a basic regulon of nitrogen metabolism was constructed. Furthermore, putative interaction partners of GlnR were identified, while the experimental validation of these candidates is ongoing. The role of AmtR in nitrogen-dependent regulation has been investigated for *M. smegmatis* and *Streptomyces avermitilis* and putative target genes have been found. The experimental validation is ongoing.

In respect to carbohydrate metabolism, the repertoire of carbohydrate uptake systems in the genus *Mycobacterium* was studied and predictions about the comparative growth potential of *M. smegmatis* and *Mycobacterium tuberculosis* on various carbon sources were provided. Furthermore, the glucose permease GlcP and the glucose kinase GlkA of *M. smegmatis* have been studied in greater detail based on comprehensive sequence and homology analyses. In a similar approach, the genome of the industrial relevant probiotic *Bifidobacterium longum* has been examined for carbohydrate uptake systems, which resulted in a detailed analysis of carbohydrate uptake systems and an overview about the growth potential of this organism on various sugars.

Also, the proteolytic potential of various *Corynebacterium* species based on all available genome sequences was compared and resulted in a more complete view on the housekeeping protease systems as well as on proteases putatively involved in pathogenicity and virulence of *Corynebacterium diphtheriae*.

3.2 Zusammenfassung

Basierend auf Genomanalysen und vergleichenden Homologieanalysen konnte das bislang weitgehend unbekannte Repertoire an Genen analysiert werden, das im Genus *Mycobacterium* an der Aufnahme und Assimilation von Stickstoff beteiligt ist. Putative Stoffwechselwege und übergeordnete regulatorische Netzwerke wurden konstruiert und einer vergleichenden Analyse unterzogen. Für den nicht-pathogenen *Mycobacterium smegmatis* konnte der Stickstoff-regulator GlnR durch einen kombinierten *in silico*, *in vitro* und *in vivo* Ansatz charakterisiert werden. Hierbei konnten direkte Zielgene verifiziert und ein Stickstoff-abhängiges Regulon konstruiert werden. Weiterhin wurden putative Interaktionspartner von GlnR identifiziert, für die die experimentelle Verifikation zur Zeit in Arbeit ist. Die Rolle von AmtR in der Stickstoff-abhängigen Regulation wurde für *M. smegmatis* und *Streptomyces avermitilis* untersucht und putative Zielgene konnten gefunden werden. Die experimentelle Validierung ist zur Zeit in Arbeit.

In Bezug auf den Kohlenhydrat Metabolismus wurde das Repertoire an Kohlenhydrat-Aufnahmesystemen im Genus *Mycobacterium* untersucht und vergleichende Voraussagen über das Wachstumspotential von *M. smegmatis* und *Mycobacterium tuberculosis* auf verschiedenen Kohlenstoffquellen getroffen. Weiterhin wurden die Glucosepermease GlcP und die Glucosekinase GlkA von *M. smegmatis* mittels vergleichender Sequenzanalyse und Homologiestudien detailliert studiert. Das Genom des industriell relevanten *Bifidobacterium longum* wurde auf Kohlenhydrat-Aufnahmesysteme untersucht. Ein kombinierter Ansatz resultierte in einer detaillierten Analyse der Kohlenhydrat Aufnahmesysteme und einer Übersicht über das Wachstumspotential dieses Organismus auf verschiedenen Kohlenstoffquellen.

Schlußendlich wurde das proteolytische Potential verschiedener Corynebakterien mithilfe verfügbarer Genomsequenzen verglichen und eine detaillierte Analyse sowohl der *housekeeping* Proteasesysteme als auch der Proteasen, die möglicherweise involviert sind in Pathogenität und Virulenz von *Corynebacterium diphtheriae*, erstellt.

4 References

Altschul, S.F. & Lipman, D.J. (1990) Protein database searches for multiple alignments. *Proc Natl Acad Sci USA* **87**: 5509-5513.

Amon, J., Bräu, T., Grimrath, A., Hänßler, E., Hasselt, K., Höller, M., Jeßberger, N., Ott, L., Szököl, J., Titgemeyer, F. & Burkovski, A. (2008a) Nitrogen control in *Mycobacterium smegmatis*: nitrogen-dependent expression of ammonium transport and assimilation proteins depends on the OmpR-type regulator GlnR. *J Bacteriol* **190**: 7108-7116.

Amon, J., Lüdke, A., Titgemeyer, F. & Burkovski, A. (2008b) General and regulatory proteolysis in corynebacteria, In: Corynebacteria: genomics and molecular biology, A. Burkovski (ed.) (Caister Academic Press, Norfolk, UK) pp. 183-201.

Amon, J., Titgemeyer, F. & Burkovski, A. (2009) A genomic view on nitrogen metabolism and nitrogen control in mycobacteria. *J Mol Microbiol Biotechnol* **17**: 20-29.

Angell, S., Lewis, C.G., Buttner, M.J. & Bibb, M.J. (1994) Glucose repression in *Streptomyces coelicolor* A3(2): a likely regulatory role for glucose kinase. *Mol Gen Genet* **244**: 135-143.

Armstrong, P. (2006) Proteases and protease inhibitors: a balance of activities in host-pathogen interaction. *Immunobiology* **211**: 263-281.

Bai, G., McCue, L.A. & McDonough, K.A. (2005) Characterization of *Mycobacterium tuberculosis* Rv3676 (CRPMt), a cyclic AMP receptor protein-like DNA binding protein. *J Bacteriol* **187**: 7795-7804.

Beckers, G., Strösser, J., Hildebrandt, U., Kalinowski, J., Farwick, M., Krämer, R. & Burkovski, A. (2005) Regulation of AmtR-controlled gene expression in *Corynebacterium glutamicum*: mechanism and characterization of the AmtR regulon. *Mol Microbiol* **58**: 580-595.

Bérdy, J. (2005) Bioactive microbial metabolites. *J Antibiot* **58**: 1-26.

Bertram, R., Schlicht, M., Mahr, K., Nothaft, H., Saier, M.H. & Titgemeyer, F. (2004). *In silico* and transcriptional analysis of carbohydrate uptake systems of *Streptomyces coelicolor* A3(2). *J Bacteriol* **186**: 1362-1373.

Bettenbrock, K., Sauter, T., Jahreis, K., Kremling, A., Lengeler, J.W. & Gilles, E.D. (2007) Analysis of the correlation between growth rates, EIIACrr phosphorylation and intracellular cAMP levels in *Escherichia coli* K-12. *J Bacteriol* **189**: 6891-6900.

Blattner, F.R., Plunkett, G. 3rd, Bloch, C.A., Perna, N.T., Burland, V., Riley, M., Collado-Vides, J., Glasner, J.D., Rode, C.K., Mayhew, G.F., Gregor, J., Davis, N.W., Kirkpatrick, H.A., Goeden, M.A., Rose, D.J., Mau, B. & Shao, Y. (1997) The complete genome sequence of *Escherichia coli* K-12. *Science* **277**: 1453-1462.

References

Bravo, A., Gill, S.S. & Soberon, M. (2007). Mode of action of *Bacillus thuringiensis* Cry and Cyt toxins and their potential for insect control. *Toxicon* **49**: 423-435.

Brückner, R. & Titgemeyer, F. (2002) Carbon catabolite repression in bacteria: choice of the carbon source and autoregulatory limitation of sugar utilization. *FEMS Microbiol Lett* **209**: 141-148.

Bueno, R., Pahel, G. & Magasanik, B. (1985) Role of *glnB* and *glnD* gene products in regulation of the *glnALG* operon of *Escherichia coli*. *J Bacteriol* **164**: 816-822.

Burkovski, A. (2007) Nitrogen control in *Corynebacterium glutamicum*: proteins, mechanisms, signals. *J Microbiol Biotechnol* **17**: 187-194.

Cerdeno-Tarraga, A.M., Efstratiou, A., Dover, L.G., Holden, M.T.G., Pallen, M., Bentley, S.D., Besra, G.S., Churcher, C., James, K.D., De Zoysa, A., Chillingworth, T., Cronin, A., Dowd, L., Feltwell, T., Hamlin, N., Holroyd, S., Jagels, K., Moule, S., Quail, M.A., Rabbinowitch, E., Rutherford, K.M., Thomson, N.R., Unwin, L., Whitehead, S., Barrell, B.G. & Parkhill, J. (2003) The complete genome sequence and analysis of *Corynebacterium diphtheriae* NCTC13129. *Nucleic Acids Res* **31**: 6516-6523.

Chaillou, S., Bor, Y.C., Batt, C.A., Postma, P.W. & Pouwels, P.H. (1998) Molecular cloning and functional expression in *Lactobacillus plantarum* 80 of *xylT* , encoding the D-xylose-H$^+$ symporter of *Lactobacillus brevis*. *Appl Environ Microbiol* **64**: 4720-4728.

Chater, K.F. (1993) Genetics of differentiation in *Streptomyces*. *Annu Rev Microbiol* **47**: 685-713.

Cole, S.T., Brosch, R., Parkhill, J., Garnier, T., Churcher, C., Harris, D., Gordon, S.V., Eiglmeier, K., Gas, S., Barry, C.E. 3rd, Tekaia, F., Badcock, K., Basham, D., Brown, D., Chillingworth, T., Connor, R., Davies, R., Devlin, K., Feltwell, T., Gentles, S., Hamlin, N., Holroyd, S., Hornsby, T., Jagels, K., Krogh, A., McLean, J., Moule, S., Murphy, L., Oliver, K., Osborne, J., Quail, M.A., Rajandream, M.A., Rogers, J., Rutter, S., Seeger, K., Skelton, J., Squares, R., Squares, S., Sulston, J.E., Taylor, K., Whitehead, S. & Barrell, B.G. (1998) Deciphering the biology of *Mycobacterium tuberculosis* from the complete genome sequence. *Nature* **393**: 537-544.

Connell, N.D., & Nikaido, H. (1994) Membrane permeability and transport in *Mycobacterium tuberculosis*. In: Tuberculosis: pathogenesis, protection, and control. B.R. Bloom (ed.) (ASM Press, Washington, DC), p. 333-349.

Content, J., Braibant, M., Connell, N. & Ainsa, J. A. (2005) Transport processes. In: Tuberculosis and the tubercle bacillus. S.T. Cole, K.D. Eisenach, D.N. McMurray & W.R. Jacobs (eds.) (ASM Press, Washington, DC), p. 379-404.

Coyle, M.B. & Lipsky, B.A. (1990) Coryneform bacteria in infectious diseases: clinical and laboratory aspects. *Clin Microbiol Rev* **3**: 227-246.

Crickmore, N. (2005) Using worms to better understand how *Bacillus thuringiensis* kills insects. *Trends Microbiol* **13**: 347-350.

References

Death, A. & Ferenci, T. (1994) Between feast and famine: endogenous inducer synthesis in the adaptation of *Escherichia coli* to growth with limiting carbohydrates. *J Bacteriol* **176**: 5101-5107.

Derouaux, A., Halici, S., Nothaft, H., Neutelings, T., Moutzourelis, G., Dusart, J., Titgemeyer, F. & Rigali, S. (2004) Deletion of a cyclic AMP receptor protein homologue diminishes germination and affects morphological development of *Streptomyces coelicolor*. *J Bacteriol* **186**: 1893-1897.

Edson, N. L. (1951) The intermediary metabolism of the mycobacteria. *Bacteriol Rev* **15**: 147-182.

Eiglmeier, K., Parkhill, J., Honore, N., Garnier, T., Tekaia, F., Telenti, A., Klatser, P., James, K.D., Thomson, N.R., Wheeler, P.R., Churcher, C., Harris, D., Mungall, K., Barrell, B.G. & Cole, S.T. (2001). The decaying genome of *Mycobacterium leprae*. *Lepr Rev* **72**: 387-398.

El Qaidi, S., Allemand, F., Oberto, J. & Plumbridge, J. (2009) Repression of *galP*, the galactose transporter in *Escherichia coli*, requires the specific regulator of N-acetylglucosamine metabolism. *Mol Microbiol* **71**: 146-157.

Embley, T.M. & Stackebrandt, E. (1994) The molecular phylogeny and systematics of the actinomycetes. *Annu Rev Microbiol* **48**: 257-289.

Engels, S., Ludwig, C., Schweitzer, J.-E., Mack, C., Bott, M. & Schaffer, S. (2005) The transcriptional regulator ClgR controls transcription of genes involved in proteolysis and DNA repair in *Corynebacterium glutamicum*. *Mol Microbiol* **57**: 576-591.

Erni, B. & Zanolari, B. (1986) Glucose-permease of the bacterial phosphotransferase system. Gene cloning, overproduction, and amino acid sequence of enzyme II^{Glc}. *J Biol Chem* **261**: 16398-16403.

Ferenci, T. (1996) Adaptation to life at micromolar nutrient levels: the regulation of *Escherichia coli* glucose transport by endoinduction and cAMP. *FEMS Microbiol Rev* **18**: 301-317.

Fink, D., Weisschuh, N., Reuther, J., Wohlleben, W. & Engels, A. (2002) Two transcriptional regulators GlnR and GlnRII are involved in regulation of nitrogen metabolism in *Streptomyces coelicolor* A3(2). *Mol Microbiol* **46**: 331-347.

Finn, R.D., Mistry, J., Tate, J., Coggill, P., Heger, A., Pollington, J.E., Gavin, O.L., Gunasekaran, P., Ceric, G., Forslund, K., Holm, L., Sonnhammer, E.L., Eddy, S.R. & Bateman, A. (2010) The Pfam protein families database. *Nucleic Acids Res* **38**: D211-222.

Flannagan, R.S., Cosío, G. & Grinstein, S. (2009) Antimicrobial mechanisms of phagocytes and bacterial evasion strategies. *Nat Rev Microbiol* **7**: 355-366.

Forchhammer, K., Hedler, A., Strobel, H. & Weiss, V. (1999) Heterotrimerization of PII-like signalling proteins: implications for PII-mediated signal transduction systems. *Mol Microbiol* **33**: 338-349.

References

Forchhammer, K. (2004) Global carbon/nitrogen control by PII signal transduction in cyanobacteria: from signals to targets. *FEMS Microbiol Rev* **28**: 319-333.

Forchhammer, K. (2007) Glutamine signalling in bacteria. *Front Biosci* **12**: 358-370.

Forchhammer, K. (2008) P(II) signal transducers: novel functional and structural insights. *Trends Microbiol* **16**: 65-72.

Franke, W. & Schillinger, A. (1944) Zum Stoffwechsel der säurefesten Bakterien I. Orientierende aerobe Reihenversuche. *Biochem Zeitung* **319**: 313-334.

Fudou, R., Jojima, Y., Seto, A., Yamada, K., Kimura, E., Nakamatsu, T., Hiraishi, A. & Yamanaka, S. (2002) *Corynebacterium efficiens* sp. nov., a glutamic-acid-producing species from soil and vegetables. *Int J Syst Evol Microbiol* **52**: 1127-1131.

Gómez-Valero, L., Rocha, E.P., Latorre, A. & Silva, F.J. (2007) Reconstructing the ancestor of *Mycobacterium leprae*: the dynamics of gene loss and genome reduction. *Genome Res* **17**: 1178-1185.

Gottesman, S. (1999) Regulation by proteolysis: developmental switches. *Curr Opin Microbiol* **2**: 142-147.

Gutknecht, R., Flukiger, K., Lanz, R. & Erni, B. (1999) Mechanism of phosphoryl transfer in the dimeric IIABMan subunit of the *Escherichia coli* mannose transporter. *J Biol Chem* **274**: 6091-6096.

Hadfield, T.L., McEvoy, P., Polotsky, Y., Tzinserling, V.A. & Yakovlev, A.A. (2000) The pathology of diphtheria. *J Infect Dis* **181**: S116-20.

Hänßler, E. & Burkovski, A. (2008) Molecular mechanisms of nitrogen control in corynebacteria. In: Corynebacteria: genomics and molecular biology. A. Burkovski (ed) (Caister Academic Press, Norfolk, UK), p. 183-201.

Hansmeier, N., Chao, T.C., Kalinowski, J., Pühler, A., & Tauch, A. (2006a) Mapping and comprehensive analysis of the extracellular and cell surface proteome of the human pathogen *Corynebacterium diphtheriae*. *Proteomics* **6**: 2465-2476.

Hansmeier, N., Chao, T.C., Pühler, A. & Tauch, A. (2006b) The cytosolic, cell surface and extracellular proteomes of the biotechnologically important soil bacterium *Corynebacterium efficiens* YS-314 in comparison to those of *Corynebacterium glutamicum* ATCC 13032. *Proteomics* **6**: 233-250.

Hansmeier, N., Chao, T.C., Daschkey, S., Müsken, M., Kalinowski, J., Pühler, A. & Tauch, A. (2007) A comprehensive proteome map of the lipid-requiring nosocomial *pathogen Corynebacterium jeikeium* K411. *Proteomics* **7**: 1076-1096.

Harth, G., Clemens, D.L. & Horwitz, A.A. (1994) Glutamine synthetase of *Mycobacterium tuberculosis*: extracellular release and characterization of its enzymatic activity. *Proc Natl Acad Sci USA* **91**: 9342-9346.

References

Harth, G., Maslesa-Galic, S., Tullius, M.V. & Horwitz, A.A. (2005) All four *Mycobacterium tuberculosis glnA* genes encode glutamine synthetase activities but only GlnA1 is abundantly expressed and essential for bacterial homeostasis. *Mol Microbiol* **58**: 1157-1172.

Hengge, R. & Bukau, B. (2003) Proteolysis in prokaryotes: protein quality control and regulatory principles. *Mol Microbiol* **49**: 1451-1462.

Herrero, A., Muro-Pastor, A.M. & Flores, E. (2001) Nitrogen control in cyanobacteria. *J Bacteriol* **183**: 411-425.

Holmes, R.K. (2000) Biology and molecular epidemiology of diphtheria toxin and the *tox* gene. *J Infect Dis* **181**: S156-167.

Hopwood, D.A. (1987) The Leeuwenhoek lecture. Towards an understanding of gene switching in *Streptomyces*, the basis of sporulation and antibiotic production. *Proc R Soc Lond B Biol Sci* **235**: 121-138.

Huang, Y.S., Ge, J., Zhang, H.M., Lei, J.Q., Zhang, X.L. & Wang, H.H. (2006) Purification and characterization of the *Mycobacterium tuberculosis* FabD2, a novel malony-CoA:AcpM transacetylase of fatty acid synthase. *Protein Expr Purif* **45**: 393-399.

Ikeda, M. & Nakagawa, S. (2003) The *Corynebacterium glutamicum* genome: features and impacts on biotechnological processes. *Appl Microbiol Biotechnol* **62**: 99-109.

Izumori, K., Ueda, Y. & Yamanaka, K. (1978) Pentose metabolism in *Mycobacterium smegmatis*: comparison of L-arabinose isomerases induced by L-arabinose and D-galactose. *J Bacteriol* **133**: 413-414.

Jahreis, K., Pimentel-Schmitt, E.F., Brückner, R. & Titgemeyer, F. (2008) Ins and outs of glucose transport systems in eubacteria. *FEMS Microbiol Rev* **32**: 891-907.

Jakoby, M., Nolden, L., Meier-Wagner, J., Krämer, R. & Burkovski, A. (2000) AmtR, a global repressor in the nitrogen regulation system of *Corynebacterium glutamicum*. *Mol Microbiol* **37**: 964-977.

Jiang, P. & Ninfa, A.J. (2009a) Reconstitution of *Escherichia coli* glutamine synthetase adenylyltransferase from N-terminal and C-terminal fragments of the enzyme. *Biochemistry* **48**: 415-423.

Jiang, P. & Ninfa, A.J. (2009b) α-ketoglutarate controls the ability of the *Escherichia coli* PII signal transduction protein to regulate the activities of NRII (NtrB) but does not control the binding of PII to NRII. *Biochemistry* **48**: 11514-11521.

Jiang, P. & Ninfa, A.J. (2009c) Sensation and signaling of α-ketoglutarate and adenylylate energy charge by the *Escherichia coli* PII signal transduction protein require cooperation of the three ligand-binding sites within the PII trimer. *Biochemistry* **48**: 11522-11531.

Jiang, P., Peliska, J.A. & Ninfa, A.J. (1998a) Enzymological characterization of the signal-transducing uridylyltransferase/uridylyl-removing enzyme (EC 2.7.7.59) of *Escherichia coli* and its interaction with the PII protein. *Biochemistry* **37**: 12782-12794.

References

Jiang, P., Peliska, J.A. & Ninfa, A.J. (1998b) Reconstitution of the signal-transduction bicyclic cascade responsible for the regulation of Ntr gene transcription in *Escherichia coli*. *Biochemistry* **37**: 12795-12801.

Jiang, P., Peliska, J.A. & Ninfa, A.J. (1998c) The regulation of *Escherichia coli* glutamine synthetase revisited: role of 2-ketoglutarate in the regulation of glutamine synthetase adenylylation state. *Biochemistry* **37**: 12802-12810.

Job, C.K. (2003) Nine-banded armadillo and leprosy research. *Indian J Pathol Microbiol* **46**: 541-550.

Kalinowski, J., Bathe, B., Bartels, D., Bischoff, N., Bott, M., Burkovski, A., Dusch, N., Eggeling, L., Eikmanns, B.J., Gaigalat, L., Goesmann, A., Hartmann, M., Huthmacher, K., Krämer, R., Linke, B., McHardy, A.C., Meyer, F., Möckel, B., Pfefferle, W., Pühler, A., Rey, D.A., Rückert, C., Rupp, O., Sahm, H., Wendisch, V.F., Wiegräbe, I. & Tauch, A. (2003) The complete *Corynebacterium glutamicum* ATCC 13032 genome sequence and its impact on the production of L-aspartate-derived amino acids and vitamins. *J Biotechnol* **104**: 5-25.

Khan, A., Akhtar, S., Ahmad, J. & Sarkar, D. (2008) Presence of a functional nitrate pathway in *Mycobacterium smegmatis*. *Microb Pathog* **44**: 71-77.

Krispin, O. & Allmansberger, R. (1998) The *Bacillus subtilis* AraE protein displays a broad substrate specificity for several different sugars. *J Bacteriol* **180**: 3250-3252.

Kunst, F., Ogasawara, N., Moszer, I., Albertini, A.M., Alloni, G., Azevedo, V., Bertero, M.G., Bessières, P., Bolotin, A., Borchert, S., Borriss, R., Boursier, L., Brans, A., Braun, M., Brignell, S.C., Bron, S., Brouillet, S., Bruschi, C.V., Caldwell, B., Capuano, V., Carter, N.M., Choi, S.K., Codani, J.J., Connerton, I.F., Danchin, A. *et al.* (1997) The complete genome sequence of the gram-positive bacterium *Bacillus subtilis*. *Nature* **390**: 249-256.

Leigh, J.A. & Dodsworth, J.A. (2007) Nitrogen regulation in bacteria and archaea. *Annu Rev Microbiol* **61**: 349-377.

Leuchtenberger, W., Huthmacher, K. & Drauz, K. (2005) Biotechnological production of amino acids and derivatives: current status and prospects. *Appl Microbiol Biotechnol* **69**: 1-8.

Martin, G.E., Seamon, K.B., Brown, F.M., Shanahan, M.F., Roberts, P.E. & Henderson, P.J. (1994) Forskolin specifically inhibits the bacterial galactose-H^+ transport protein, GalP. *J Biol Chem* **269**: 24870-24877.

McCue, L.A., McDonough, K.A. & Lawrence, C.E. (2000) Functional classification of cNMP-binding proteins and nucleotide cyclases with implications for novel regulatory pathways in *Mycobacterium tuberculosis*. *Genome Res* **10**: 204-219.

Merrick, M.J. & Edwards, R.A. (1995) Nitrogen control in bacteria. *Microbiol Rev* **59**: 604-622.

Mogk, A., Schmidt, R. & Bukau, B. (2007) The N-end rule pathway for regulated proteolysis: prokaryotic and eukaryotic strategies. *Trends Cell Biol* **17**: 165-172.

References

Monedero, V., Maze, A., Boel, G., Zuniga, M., Beaufils, S., Hartke, A. & Deutscher, J. (2007) The phosphotransferase system of *Lactobacillus casei*: regulation of carbon metabolism and connection to cold shock response. *J Mol Microbiol Biotechnol* **12**: 20-32.

Moon, M.W., Park, S.Y., Choi, S.K. & Lee, J.K. (2007) The phosphotransferase system of *Corynebacterium glutamicum*: features of sugar transport and carbon regulation. *J Mol Microbiol Biotechnol* **12**: 43-50.

Muhl, D., Jeßberger, N., Hasselt, K., Jardin, C., Sticht, H. & Burkovski, A. (2009) DNA binding by *Corynebacterium glutamicum* TetR-type transcription regulator AmtR. *BMC Mol Biol* **10**: 73.

Muro-Pastor, M.I., Reyes, J.C. & Florencio, F.J. (2005) Ammonium assimilation in cyanobacteria. *Photosynth Res* **83**: 135-150.

Nolden, L., Beckers, G. & Burkovski, A (2002) Nitrogen assimilation in *Corynebacterium diphtheriae*: pathways and regulatory cascades. *FEMS Microbiol Lett* **208**: 287-293.

Osanai, T. & Tanaka, K. (2007) Keeping in touch with PII: PII-interacting proteins in unicellular cyanobacteria. *Plant Cell Physiol* **48**: 908-914.

Pao, S.S., Paulsen, I.T. & Saier, M.H. jr. (1998) Major facilitator superfamily. *Microbiol Mol Biol Rev* **62**: 1-34.

Parche, S., Burkovski, A., Sprenger, G.A., Weil, B., Kramer, R. & Titgemeyer, F. (2001a) *Corynebacterium glutamicum*: a dissection of the PTS. *J Mol Microbiol Biotechnol* **3**: 423-428.

Parche, S., Thomae, A.W., Schlicht, M. & Titgemeyer, F. (2001b) *Corynebacterium diphtheriae*: a PTS view to the genome. *J Mol Microbiol Biotechnol* **3**: 415-422.

Parche, S., Beleut, M., Rezzonico, E., Jacobs, D., Arigoni, F., Titgemeyer, F. & Jankovic, I. (2006) Lactose-over-glucose preference in *Bifidobacterium longum* NCC2705: *glcP*, encoding a glucose transporter, is subject to lactose repression. *J Bacteriol* **188**: 1260-1265.

Parche, S., Amon, J., Jankovic, I., Rezzonico, E., Beleut, M., Barutçu, H., Schendel, I., Eddy, M.P., Burkovski, A., Arigoni, F. & Titgemeyer, F. (2007) Sugar transport systems of *Bifidobacterium longum* NCC2705. *J Mol Microbiol Biotechnol* **12**: 9-19.

Parish, T. & Stoker, N. G. (2000) *glnE* is an essential gene in *Mycobacterium tuberculosis*. *J Bacteriol* **182**: 5715-5720.

Park, Y.H., Lee, B.R., Seok, Y.J. & Peterkofsky, A. (2006) In vitro reconstitution of catabolite repression in *Escherichia coli*. *J Biol Chem* **281**: 6448–6454.

Pimentel-Schmitt, E.F., Jahreis, K., Eddy, M.P., Amon, J., Burkovski, A. & Titgemeyer, F. (2009) Identification of a glucose permease from *Mycobacterium smegmatis* mc2 155. *J Mol Microbiol Biotechnol* **16**: 169-175.

References

Pimentel-Schmitt, E.F., Thomae, A.W., Amon, J., Klieber, M.A., Roth, H.M., Muller, Y.A., Jahreis, K., Burkovski, A. & Titgemeyer, F. (2007) A glucose kinase from *Mycobacterium smegmatis*. *J Mol Microbiol Biotechnol* **12**: 75-81.

Postma, P.W., Lengeler, J.W. & Jacobson, G.R. (1993) Phosphoenolpyruvate:carbohydrate phosphotransferase systems of bacteria. *Microbiol Rev* **57**: 543–594.

Ranque, B., Alter, A., Schurr, E., Abel, L. & Alcais, A.(2008) Leprosy: a paradigm for the study of human genetic susceptibility to infectious diseases. *Med Sci* **24**: 491-497.

Rao, M.B., Tanksale, A.M., Ghatge, M.S. & Deshpande, V.V. (1998) Molecular and biotechnological aspects of microbial proteases. *Microbiol Mol Biol Rev* **62**: 597-635.

Ratledge, C. & Stanford, J. (1982) Nutrition, growth, and metabolism. In: The biology of mycobacteria, C. Ratledge and J. Stanford (eds.) (Academic Press Inc. Ltd., London, United Kingdom), p.186-271.

Rawlings, N.D., Morton, F.R. & Barrett, A.J. (2006) MEROPS: the peptidase database. *Nucleic Acids Res* **34**: D270-272.

Read, R., Pashley, C. A., Smith, D. & Parish, T. (2007) The role of GlnD in ammonium assimilation in *Mycobacterium tuberculosis*. *Tuberculosis* **87**: 384-390.

Reizer, J., Hoischen, C., Titgemeyer, F., Rivolta, C., Rabus, R., Stülke, J., Karamata, D., Saier, M.H. Jr., & Hillen, W. (1998) A novel protein kinase that controls carbon catabolite repression in bacteria. *Mol Microbiol* **27**: 1157-1169.

Reuther, J. & Wohlleben, W. (2007) Nitrogen metabolism in *Streptomyces coelicolor*: transcriptional and post-translational regulation. *J Mol Microbiol Biotechnol* **12**: 139-146.

Ribeiro-Guimaraes, M.L. & Pessolani, M.C. (2007) Comparative genomics of mycobacterial proteases. *Microb Pathog* **43**: 5-6.

Ribeiro-Guimaraes, M.L., Tempone, A.J., Amaral, J.J., Nery, J.A., Antunes, S.L. & Pessolani, M.C. (2007) Expression analysis of proteases of *Mycobacterium leprae* in human skin lesions. *Microb Pathog* **43**: 249-254.

Rigali, S., Nothaft, H., Noens, E.E., Schlicht, M., Colson, S., Müller, M., Joris, B., Körten, H.K., Hopwood, D.A., Titgemeyer, F. &van Wezel, G.P. (2006) The sugar phosphotransferase system of *Streptomyces coelicolor* is regulated by the GntR-family regulator DasR and links N-acetylglucosamine metabolism to the control of development. *Mol Microbiol* **61**: 1237-1251.

Romano, A.H., Eberhard, S.J., Dingle, S.L. & McDowell, T.D. (1970) Distribution of the phosphoenolpyruvate:glucose phosphotransferase system in bacteria. *J Bacteriol* **104**: 808-813.

Ronaghi, M., Uhlén, M. & Nyrén, P. (1998) A sequencing method based on real-time pyrophosphate. *Science* **281**: 363, 365.

References

Saier, M.H. Jr., Tran, C.V. & Barabote, R.D. (2006) TCDB: the Transporter Classification Database for membrane transport protein analyses and information. *Nucleic Acids Res* **34**: D181-186.

Sanger, F. & Coulson, A.R. (1975) A rapid method for determining sequences in DNA by primed synthesis with DNA polymerase. *J Mol Biol* **94**: 441-448.

Sassetti, C.M., Boyd, D.H. & Rubin, E.J. (2003) Genes required for mycobacterial growth defined by high density mutagenesis. *Mol Microbiol* **48**: 77-84.

Sassetti, C.M. & Rubin, E.J. (2003) Genetic requirements for mycobacterial survival during infection.
Proc Natl Acad Sci USA **100**: 12989-12994.

Schallmey, M., Singh, A. & Ward, O.P. (2004) Developments in the use of *Bacillus* species for industrial production. *Can J Microbiol* **50**: 1-17.

Schell, M.A., Karmirantzou, M., Snel, B., Vilanova, D., Berger, B., Pessi, G., Zwahlen, M.C., Desiere, F., Bork, P., Delley, M., Pridmore, R.D. & Arigoni, F. (2002) The genome sequence of *Bifidobacterium longum* reflects its adaptation to the human gastrointestinal tract. *Proc Natl Acad Sci USA* **99**: 14422-14427.

Schrempf, H. (2001) Recognition and degradation of chitin by streptomycetes. *Antonie van Leuwenhook* **79**: 285-289.

Shawver, L.K., Slamon, D. & Ullrich, A. (2002) Smart drugs: tyrosine kinase inhibitors in cancer therapy. *Cancer Cell* **1**: 117-123.

Sohaskey, C.D. (2008) Nitrate enhances the survival of *Mycobacterium tuberculosis* during inhibition
of respiration. *J Bacteriol* **190**: 2981-2986.

Strösser, J., Lüdke, A., Schaffer, S., Krämer, R. & Burkovski, A. (2004) Regulation of GlnK activity: modification, membrane sequestration, and proteolysis as regulatory principles in the network of nitrogen control in *Corynebacterium glutamicum*. *Mol Microbiol* **54**: 132-147.

Tauch, A., Kaiser, O., Hain, T., Goesmann, A., Weisshaar, B., Albersmeier, A., Bekel, T., Bischoff, N., Brune, I., Chakraborty, T., Kalinowski, J., Meyer, F., Rupp, O., Schneiker, S., Viehoever, P. & Pühler, A. (2005) Complete genome sequence and analysis of the multiresistant nosocomial pathogen *Corynebacterium jeikeium* K411, a lipid-requiring bacterium of the human skin flora. *J Bacteriol* **187**: 4671-4682.

Thanbichler, M., Wang, S.C. & Shapiro, L. (2005) The bacterial nucleoid: a highly organized and dynamic structure. *J Cell Biochem* **96**: 506-521.

Tiffert, Y., Supra, P., Wurm, R., Wohlleben, W., Wagner, R. & Reuther, J. (2008) The *Streptomyces coelicolor* GlnR regulon: identification of new GlnR targets and evidence for a central role of GlnR in nitrogen metabolism in actinomycetes. *Mol Microbiol* **67**: 861-880.

References

Titgemeyer, F., Reizer, J., Reizer, A. & Saier, M.H. Jr. (1994a) Evolutionary relationships between sugar kinases and transcriptional repressors in bacteria. *Microbiology* **140**: 2349-2354.

Titgemeyer, F., Walkenhorst, J., Cui, X., Reizer, J. & Saier, M.H. Jr. (1994b) Proteins of the phosphoenolpyruvate:sugar phosphotransferase system in *Streptomyces* : possible involvement in the regulation of antibiotic production. *Res Microbiol* **145**: 89-92.

Titgemeyer, F., Amon, J., Parche, S., Mahfoud, M., Bail, J., Schlicht, M., Rehm, N., Hillmann, D., Stephan, J., Walter, B., Burkovski, A. & Niederweis, M. (2007) A genomic view of sugar transport in *Mycobacterium smegmatis* and *Mycobacterium tuberculosis*. *J Bacteriol* **189**: 5903-5915.

Tullius, M V., Harth, G. & Horwitz, A.A. (2003) Glutamine synthetase GlnA1 is essential for growth of *Mycobacterium tuberculosis* in human THP-1 macrophages and guinea pigs. *Infect Immun* **71**: 3927-3936.

van Heeswijk, W.C., Stegeman, B., Hoving, S., Molenaar, D., Kahn, D. & Westerhoff, H.V. (1995) An additional PII in *Escherichia coli*: a new regulatory protein in the glutamine synthetase cascade. *FEMS Microbiol Lett* **132**: 153-157.

van Heeswijk, W.C., Hoving, S., Molenaar, D., Stegeman, B., Kahn, D. & Westerhoff, H.V. (1996) An alternative PII protein in the regulation of glutamine synthetase in *Escherichia coli*. *Mol Microbiol* **21**: 133-146.

van Heeswijk, W.C., Wen, D., Clancy, P., Jaggi, R., Ollis, D.L., Westerhoff, H.V. & Vasudevan, S.G. (2000) The *Escherichia coli* signal transducers PII (GlnB) and GlnK form heterotrimers in vivo: fine tuning the nitrogen signal cascade. *Proc Natl Acad Sci USA* **97**: 3942-3947.

van Heeswijk, W.C., Molenaar, D., Hoving, S. & Westerhoff, H.V. (2009) The pivotal regulator GlnB of *Escherichia coli* is engaged in subtle and context-dependent control. *FEBS J* **276**: 3324-3340.

van Roosmalen, M.L., Geukens, N., Jongbloed, J.D.H., Tjalsma, H., Dubois, J.-Y.F., Bron, S., van Dijl, J.M. & Anné, J. (2004) Type I signal peptidases of Gram-positive bacteria. *Biochim Biophys Acta* **1694**: 279-297.

van Wezel, G.P., Mahr, K., König, M., Traag, B.A., Pimentel-Schmitt, E.F., Willimek, A. & Titgemeyer, F. (2005) GlcP constitutes the major glucose uptake system of *Streptomyces coelicolor* A3(2). *Mol Microbiol* **55**: 624-636.

van Wezel, G.P., König, M., Mahr, K., Nothaft, H., Thomae, A.W., Bibb, M.J. & Titgemeyer, F. (2007) A new piece of an old jigsaw: glucose kinase is activated posttranslationally in a glucose transport-dependent manner in *Streptomyces coelicolor* A3 (2). *J Mol Microbiol Biotechnol* **12**: 65-72.

Ventura, M., Canchaya, C., Fitzgerald, G.F., Gupta, R.S. & van Sinderen, D. (2007a) Genomics as a means to understand bacterial phylogeny and ecological adaptation: the case of bifidobacteria. *Antonie van Leeuwenhoek* **91**: 351-372.

Ventura, M., Canchaya, C., Tauch, A., Chandra, G., Fitzgerald, G.F., Chater, K.F. & van Sinderen, D. (2007b) Genomics of Actinobacteria: tracing the evolutionary history of an ancient phylum. *Microbiol Mol Biol Rev* **71:** 495-548.

Vissa, V.D. & Brennan, P.J. (2001) The genome of *Mycobacterium leprae*: a minimal mycobacterial genome set. *Genome Biol* **2:** 1023.1-1023.8.

Walter, B., Hänßler, E., Kalinowski, J. & Burkovski, A. (2007) Nitrogen metabolism and nitrogen control in corynebacteria: variations of a common theme. *J Mol Microbiol Biotechnol* **12:** 131-138.

Weickert, M.J. & Adhya, S. (1993) The galactose regulon of *Escherichia coli*. *Mol Microbiol* **10:** 245-251.

Weiss, V., Kramer, G., Dünnebier, T. & Flotho, A. (2002) Mechanism of regulation of the bifunctional histidine kinase NtrB in *Escherichia coli*. *J Mol Microbiol Biotechnol* **4:** 229-233.

Westers, L., Westers, H. & Quax, W.J. (2004) *Bacillus subtilis* as cell factory for pharmaceutical proteins: a biotechnological approach to optimize the host organism. *Biochim Biophys Acta* **1694:** 299-310.

Wickner, S., Maurizi, M.R. & Gottesman, S. (1999). Posttranslational quality control: folding, refolding, and degrading proteins. *Science* **286:** 1888-1893.

Wray, L.V. & Fisher, S.H. (1993) The *Streptomyces coelicolor* glnR gene encodes a protein similar to other bacterial response regulators. *Gene* **130:** 145-150.

Wray, L.V., Atkinson, M.R. & Fisher, S.H. (1991) Identification and cloning of the *glnR* locus, which is required for transcription of the *glnA* gene in *Streptomyces coelicolor* A3(2). *J Bacteriol* **173:** 7351-7360.

Yoshida, T., Qin, L., Egger, L.A. & Inouye, M. (2006) Transcription regulation of *ompF* and *ompC* by a single transcription factor, OmpR. *J Biol Chem* **281:** 17114-17123.

Yu, H., Peng, W., Liu, Y., Wu, T., Yao, Y., Cui, M., Jiang, W. & Zhao, G.-P. (2006) Identification and characterization of *glnA* promoter and its corresponding trans-regulating protein GlnR in the rifamycin SV producing actinomycete, *Amycolatopsis mediterranei* U32. *Acta Biochim Biophys Sin* **38:** 831-843.

Yu, H., Yao, Y., Liu, Y., Jiao, R., Jiang, W. & Zhao, G.-P. (2007) A complex role of *Amycolatopsis mediterranei* GlnR in nitrogen metabolism and related antibiotics production. *Arch Microbiol* **188:** 89-96.

Zhang, C.C., Durand, M.C., Jeanjean, R. & Joset, F. (1989) Molecular and genetical analysis of the fructose-glucose transport system in the cyanobacterium *Synechocystis* PCC6803. *Mol Microbiol* **3:** 1221-1229.

5 Own publications

Amon, J., Bräu, T., Grimrath, A., Hänßler, E., Hasselt, K., Höller, M., Jeßberger, N., Ott, L., Szököl, J., Titgemeyer, F. & Burkovski, A. (2008a) Nitrogen control in *Mycobacterium smegmatis*: nitrogen-dependent expression of ammonium transport and assimilation proteins depends on the OmpR-type regulator GlnR. *J Bacteriol* **190**: 7108-7116.

Amon, J., Lüdke, A., Titgemeyer, F, & Burkovski, A. (2008b) General and regulatory proteolysis in corynebacteria, In: Corynebacteria: genomics and molecular biology, A. Burkovski (ed.) (Caister Academic Press, Norfolk, UK) pp. 183-201.

Amon, J., Titgemeyer, F. & Burkovski, A. (2009) A genomic view on nitrogen metabolism and nitrogen control in mycobacteria. *J Mol Microbiol Biotechnol* **17**: 20-29.

Parche, S., **Amon, J.**, Jankovic, I., Rezzonico, E., Beleut, M., Barutçu, H., Schendel, I., Eddy, M.P., Burkovski, A., Arigoni, F. & Titgemeyer, F. (2007) Sugar transport systems of *Bifidobacterium longum* NCC2705. *J Mol Microbiol Biotechnol* **12**: 9-19.

Pimentel-Schmitt, E.F., Jahreis, K., Eddy, M.P., **Amon, J.**, Burkovski, A. & Titgemeyer, F. (2009) Identification of a glucose permease from *Mycobacterium smegmatis* mc2 155. *J Mol Microbiol Biotechnol* **16**: 169-175.

Pimentel-Schmitt, E.F., Thomae, A.W., **Amon, J.**, Klieber, M.A., Roth, H.M., Muller, Y.A., Jahreis, K., Burkovski, A. & Titgemeyer, F. (2007) A glucose kinase from *Mycobacterium smegmatis*. *J Mol Microbiol Biotechnol* **12**: 75-81.

Titgemeyer, F., **Amon, J.**, Parche, S., Mahfoud, M., Bail, J., Schlicht, M., Rehm, N., Hillmann, D., Stephan, J., Walter, B., Burkovski, A. & Niederweis, M. (2007) A genomic view of sugar transport in *Mycobacterium smegmatis* and *Mycobacterium tuberculosis*. *J Bacteriol* **189**: 5903-5915.

6 Appendix

6.1 Full species names

Table 3: Full taxonomic names of species used throughout the manuscript.

Actinobacteria:
Acidothermus cellulolyticus
Actinomyces naeslundii
Amycolatopsis mediterranei
Arthrobacter aurescens
Bifidobacterium longum
Catenulispora acidiphila
Clavibacter michiganensis ssp. *michiganensis*
Clavibacter michiganensis ssp. *sepedonicus*
Corynebacterium accolens
Corynebacterium aurimucosum
Corynebacterium diphtheria
Corynebacterium efficiens
Corynebacterium glutamicum
Corynebacterium lipophiloflavum
Corynebacterium matruchotii
Corynebacterium pseudogenitalium
Corynebacterium striatum
Gardnerella vaginalis
Geodermatophilus obscurus
Gordonia bronchialis
Kineococcus radiotolerans
Kocuria rhizophila
Kribbella flavida
Lactobacillus brevis
Leifsonia xyli
Mycobacterium abscessus

Appendix

Mycobacterium avium paratuberculosis
Mycobacterium bovis
Mycobacterium smegmatis
Mycobacterium tuberculosis
Nakamurella multipartita
Nocardia farcinica
Propionibacterium acnes
Renibacterium salmoninarum
Rhodococcus erythropolis
Rhodococcus jostii
Rhodococcus opacus
Salinispora arenicola
Streptomyces avermitilis
Streptomyces coelicolor
Streptomyces griseus
Streptosporangium roseum
Thermobifida fusca
Thermonospora curvata
Tsukamurella paurometabola
Zymomonas mobilis

6.2 Publications

Research Article

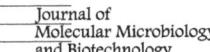
J Mol Microbiol Biotechnol
DOI: 10.1159/000159195

Published online: September 29, 2008

A Genomic View on Nitrogen Metabolism and Nitrogen Control in Mycobacteria

Johannes Amon   Fritz Titgemeyer   Andreas Burkovski

Lehrstuhl für Mikrobiologie, Friedrich-Alexander-Universität Erlangen-Nürnberg, Erlangen, Deutschland

Key Words
Ammonium · *Mycobacterium* · Glutamine synthetase · Nitrogen control · Nitrogen regulation · GlnR

Abstract
Knowledge about nitrogen metabolism and control in the genus *Mycobacterium* is sparse, especially compared to the state of knowledge in related actinomycetes like *Streptomyces coelicolor* or the close relative *Corynebacterium glutamicum*. Therefore, we screened the published genome sequences of *Mycobacterium smegmatis*, *Mycobacterium tuberculosis*, *Mycobacterium bovis*, *Mycobacterium avium* ssp. *paratuberculosis* and *Mycobacterium leprae* for genes encoding proteins for uptake of nitrogen sources, nitrogen assimilation and nitrogen control systems, resulting in a detailed comparative genomic analysis of nitrogen metabolism-related genes for all completely sequenced members of the genus. Transporters for ammonium, nitrate, and urea could be identified, as well as enzymes crucial for assimilation of these nitrogen sources, i.e. glutamine synthetase, glutamate dehydrogenase, glutamate synthase, nitrate reductase, nitrite reductase, and urease proteins. A reduction of genes encoding proteins for nitrogen transport and metabolism was observed for the pathogenic mycobacteria, especially for *M. leprae*. Signal transduction components identified for the different species include adenylyl- and uridylyltransferase and a P_{II}-type signal transduction protein. Exclusively for *M. smegmatis*, two homologs of putative nitrogen regulatory proteins were found, namely GlnR and AmtR, while in other mycobacteria, AmtR was absent and GlnR seems to be the nitrogen transcription regulator protein.

Copyright © 2008 S. Karger AG, Basel

Introduction

The genus *Mycobacterium* includes saprophytic soil bacteria like *Mycobacterium smegmatis* as well as human pathogens such as *Mycobacterium leprae* and, the most prominent member, *Mycobacterium tuberculosis*. The latter is responsible for an estimated eight million new infections and nearly two million deaths each year [Dye et al., 2005]. Due to the ability of *M. tuberculosis* to replicate within macrophages, these bacteria are able to escape the human immune system and can persist in the human lung for decades in a state of dormancy prior to reactivation. Medical treatment of tuberculosis is further impeded by the appearance of strains, which are highly resistant against almost all standard antibiotics. For the design of new drugs, a more thorough understanding of the molecular biology and physiology of mycobacteria becomes crucial. A milestone in this respect has been achieved by sequencing several mycobacterial genomes including *M. tuberculosis*, *M. leprae*, *Mycobacterium avium* and *Mycobacterium bovis* [Cole et al., 1998, 2001; Garnier et al., 2003; Li et al., 2005]. This is also the case for the fast-growing *M. smegmatis* (http://www.tigr.org/), which has been used as the basis of mycobacterial genetics [Pelicic et al., 1998] and as a non-pathogen substitute to study the metabolic and regulatory pathways of *M. tuberculosis* [Bange et al., 1999; Titgemeyer et al., 2007].

We are especially interested in nitrogen metabolism and regulation in mycolic acids-containing actinomycetes. Since almost all of the macromolecules in a bacterial cell, e.g. proteins, nucleic acids and cell wall components, contain nitrogen, prokaryotes have developed elaborate mechanisms to provide an optimal nitrogen supply for cell maintenance and growth, and to overcome and survive situations of nitrogen starvation. In actinomycetes, the uptake and assimilation of nitrogen sources has been studied intensively for the close relatives of mycobacteria *Corynebacterium glutamicum* [for reviews, see Burkovski, 2003a, b, 2005, 2007; Hänßler and Burkovski, 2008] and *Streptomyces coelicolor* [Reuther and Wohlleben, 2007], while for the genus *Mycobacterium*, work concentrated mainly on the glutamine synthetase GlnA1 and its adenylylase GlnE as potential drug targets [Carroll et al., 2008; Harth et al., 2005; Hotter et al., 2008; Nordqvist et al., 2008]. Using the information available for these well-studied model organisms, we compiled a list of nitrogen metabolism-related genes based on in silico analyses of the published genome sequence of *M. smegmatis* mc^{2}155 and their respective homologs in various pathogenic mycobacteria to provide comprehensive information for further experimental studies.

Results and Discussion

Genes Related to Ammonium Uptake and Assimilation

Ammonium is the preferred nitrogen source of most microorganisms and consequently ammonium uptake systems and assimilating enzymes are widely distributed among bacteria. A highly conserved apparent operon was found in all mycobacterial genomes comprising the genes *amtB-glnK-glnD*. The *amtB* gene encodes a putative ammonium uptake system, which has been functionally characterized in enterobacteria, *Bacillus subtilis* [Detsch and Stülke, 2003] and *C. glutamicum* [Jakoby et al., 1999; Meier-Wagner et al., 2001; Walter et al., 2008]. The two other genes are crucial for nitrogen-dependent signal transduction. The *glnK* gene is coding for a P$_{II}$-type signal transduction protein and *glnD* encodes a GlnK modifying/demodifying enzyme, which works as a uridylyltransferase in enterobacteria and acts as an adenylyltransferase in *S. coelicolor* [Hesketh et al., 2002] and *C.* *glutamicum* [Strösser et al., 2004]. In accordance with the massive gene decay observed in the leprosy bacillus, in *M. leprae* all genes of this operon are pseudogenes and carry mutations, which prevent synthesis of active proteins. For example, *amtB* exhibits a mutated start codon, thus preventing translation, and multiple frame shifts resulting in a nonfunctional reading frame.

In *M. smegmatis*, another two genes, designated *amtA* (*msmeg_4635*) and *amt1* (*msmeg_6259*) were found, which encode additional putative ammonium transporters. While for *amtA* no apparent operon was observed, the latter one is organized in a cluster together with genes that apparently encode different functional domains of a class III glutamine synthetase (*msmeg_6260*), a class II glutamine amidotransferase (*msmeg_6261*), and domains of a glutamate synthase (*msmeg_6262*, *msmeg_6263*). This conserved region seems to be of Pseudomonadales origin according to the BLAST results and is in the sequenced actinomycetes only found in the genomes of *M. smegmatis*, *Nocardia farcinica*, and *Rhodococcus* sp.

Once ammonium has entered the cell via diffusion across the cytoplasmic membrane or by protein-dependent transport, this nitrogen source has to be assimilated. Most bacteria have two primary pathways to facilitate the incorporation of ammonium into the key nitrogen donors for biosynthetic reactions, L-glutamate and L-glutamine, namely the glutamate dehydrogenase (GDH) and the glutamine synthetase/glutamate synthase pathways (GS/GOGAT). Assimilation via glutamate dehydrogenase is bioenergetically more favorable, as the GS/GOGAT pathway utilizes an additional ATP per assimilated ammonium. Consequently, GDH is often preferentially used in ammonium-rich medium. However, when the cells face nitrogen limitation, assimilation via GDH is not sufficient due to the low affinity of the enzyme and the GS/GOGAT pathway is recruited for ammonium assimilation. Interestingly, only the genome of *M. smegmatis* features a *gdhA* gene coding for an assimilatory NADPH-dependent glutamate dehydrogenase. This gene is missing in the other four mycobacterial species investigated here and consequently ammonium assimilation depends on the GS/GOGAT pathway in these bacteria.

The genes *glnA1* and *glnA2* are found in all mycobacterial genomes in a conserved region together with *glnE*, encoding an adenylyltransferase that is essential in *M. tuberculosis* [Parish et al., 2000] and was shown to regulate glutamine synthetase activity [Carroll et al., 2008]. Homologs of the GlnA3 and GlnA4 proteins were found in all mycobacterial genomes except in *M. leprae*, while for GlnA4 at least three homologous proteins were identified in the genome of *M. smegmatis* via BLASTP (MSMEG_2595, MSMEG_3828, MSMEG_1116), of which MSMEG_2595 shares the highest identity to the *S. coelicolor* GlnA4 protein (62%, SCO1613 [Rexer et al., 2006]). Besides homologs of *glnA1*, *glnA2*, *glnA3* and *glnA4*, *M. smegmatis* possesses various additional genes that seem to encode homologs of glutamine synthetases (see below and table 1), and which are missing in other mycobacterial genomes. MSMEG_5374 shares with 51% the highest identity to glutamine synthetases-like proteins in the soil-dwelling α proteobacteria *Rhodopseudomonas palustris* CGA009 (RPA0984) and *Bradyrhizobium japonicum* USDA 110 (BLL1069). Furthermore, *msmeg_5374* features with 63% a GC-content significantly below the average 67% G+C content of the genome of *M. smegmatis*. For MSMEG_3827, only weak identities (about 30%) to various not further characterized glutamine synthetase-like proteins in proteobacteria were found. MSMEG_6260, which also shares 69% identity to a putative glutamine synthetase of *N. farcinica* (NFA21040), seems to be a type III glutamine synthetase with around 45% identity to GS III enzymes of various *Synechococcus* and *Pseudomonas* species, also indicated by the BLASTP results of the corresponding putative operon *(msmeg_6260-6264)*. Another putative type III GS might be encoded by *msmeg_6693*, exhibiting 35–45% identical amino acids to various rhizobial GS III enzymes and 33% identity to the glutamine synthetase III of *Agrobacterium tumefaciens* (Atu4230). While it was already shown for *M. tuberculosis* that only GlnA1 is abundantly expressed and essential for bacterial homeostasis [Harth et al., 2005], the physiological role and function of the various other glutamine synthetase-encoding genes, especially in the genome of *M. smegmatis*, remain to be verified.

In all mycobacterial genomes, a highly conserved apparent operon encoding the large and small subunit of the glutamate synthase (GOGAT; *gltB* and *gltD*) was observed. In addition, the genome of *M. smegmatis* features several additional copies of *gltB (msmeg_5594, msmeg_6263, msmeg6459)* and *gltD (msmeg_6262, msmeg_6458)* that are not found in other mycobacteria.

Genes Encoding Transporters of Alternative Nitrogen Sources and Assimilatory Enzymes

Ammonium cannot only be provided directly, but also generated from alternative nitrogen sources. A prominent example is urea, which is degraded to ammonium and carbon dioxide by ureases. For *M. smegmatis*, two putative urease-encoding operons were found, of which only one *(msmeg_3623-3627)* exhibits homology to the *ure* gene clusters in *M. tuberculosis* and *M. bovis* based on gene identity and arrangement. The second one exhibits striking similarities to the urease subunits – encoding genes from α proteobacteria (e.g. *Pseudomonas syringae*, *Helicobacter pylori*, and various *Burkholderia* species) with DNA sequence identities between 60 and 70%, an identical operon arrangement and a *ureAB* fusion gene. No urease operon or urease-related genes were found in *M. avium* and *M. leprae*. Also, *M. smegmatis* is the only mycobacterial species to feature a distinct operon *(msmeg_2978-2982)* encoding the subunits of a putative urea ABC transporter, emphasizing – together with the presence of the different urease-encoding operons – the importance of urea as a nitrogen source for this species.

Furthermore, we found genes encoding the necessary components for the complete reduction of nitrate to ammonium via nitrate reductase (NarGHJI, NarX) and nitrite reductase (NirBD), including putative nitrite/nitrate transporters (homologs of *Escherichia coli* NarK and NarU), in the genomes of all screened mycobacteria except *M. leprae*, which seems to possess a nitrate/nitrite antiporter (ML0844) but is missing genes encoding the corresponding enzymes for reduction and assimilation.

Signal Transduction

Among mycobacteria, proteins involved in signalling and posttranslational modification are best characterized in *M. tuberculosis* [Parish and Stoker, 2000]. Since glutamine synthetase in *M. tuberculosis* is essential for this bacterium and consequently an important drug target [Harth et al., 1994, 2005; Nordqvist et al., 2008; Tullius et al., 2003], work concentrated first on the *glnE* gene product adenylyltransferase, which is involved in posttranslational modification and regulation of this enzyme. It was shown that *glnE* is essential in this organism [Parish and Stoker, 2000], an observation that is in agreement with the crucial function of adenylyltransferase in regulation of GS activity [Carroll et al., 2008]; furthermore, the *M. tuberculosis glnE* promoter is upregulated in ammonia- or glutamine-containing media, at least in the heterologous host *M. smegmatis* [Pashley et al., 2006], while the transcriptional organization of the *glnA1-glnE-glnA2* gene cluster remains to be elucidated further [Hotter et al., 2008].

GlnD seems to have no crucial function in the regulation of ammonium assimilation in *M. tuberculosis* [Read et al., 2007]. This is in agreement with observations made in *S. coelicolor* and *C. glutamicum* [Hesketh et al., 2002; Strösser et al., 2004]. In contrast to the signal transfer *via* GlnD, GlnK and GlnE to GS as shown in *E. coli*, enzyme

Table 1. Nitrogen metabolism-related genes in mycobacteria: the given references were selected to provide information for representative, well-studied homologs

Enzyme/function	Gene name	M. smegmatis mc² 155	M. tuberculosis H37rv	M. leprae TN	M. avium ssp. paratuberculosis	M. bovis ssp. bovis AF2122/97	Reference/nearest homologs
Assimilatory nitrite reductase	nirB	msmeg_0427	Rv0252	n/a	MAP_3702	Mb0258	E. coli nitrite reductase operon [Wang and Gunsalus, 2000]
	nirD	msmeg_0428	Rv0253	n/a	MAP_3703	Mb0259	
Hydrolysis of urea	ureE	msmeg_1091	n/a	n/a	n/a	n/a	Pseudomonas syringae urease operon (psyr_2200-2195)
	ureF	msmeg_1092	n/a	n/a	n/a	n/a	
	ureAB	msmeg_1093	n/a	n/a	n/a	n/a	
	ureC	msmeg_1094	n/a	n/a	n/a	n/a	
	ureG	msmeg_1095	n/a	n/a	n/a	n/a	
	ureD	msmeg_1096	n/a	n/a	n/a	n/a	
Unknown	glnA*	msmeg_1116	n/a	n/a	n/a	n/a	[Harth et al., 2005]
Ammonium uptake	amtB	msmeg_2425	Rv2920c	(ML1627)	MAP_2988c	Mb2944c	amtB-glnK-glnD S. coelicolor [Fink et al., 2002]
Signal transduction	glnK	msmeg_2426	Rv2919c	(ML1626)	MAP_2987c	Mb2943c	
Post-translational regulation of GlnK	glnD	msmeg_2427	Rv2918c	(ML1625)	MAP_2986c	Mb2942c	
Unknown	glnA4	msmeg_2595	Rv2860c	n/a	MAP_2931c	Mb2885c	[Harth et al., 2005]
Urea uptake	urtE	msmeg_2978	n/a	n/a	n/a	n/a	C. glutamicum urease transport operon [Beckers et al., 2004]
	urtD	msmeg_2979	n/a	n/a	n/a	n/a	
	urtC	msmeg_2980	n/a	n/a	n/a	n/a	
	urtB	msmeg_2981	n/a	n/a	n/a	n/a	
	urtA	msmeg_2982	n/a	n/a	n/a	n/a	
Ammonium assimilaton	gltB	msmeg_3225	Rv3859	ML0061	MAP_0172	Mb3888c	C. glutamicum glutamate synthase [Beckers et al., 2001; Schulz et al., 2001]
	gltD	msmeg_3226	Rv3858	ML0062	MAP_0173	Mb3889c	
Unknown	glnA3	msmeg_3561	Rv1878	n/a	MAP_1599	Mb1910	[Harth et al., 2005]
Hydrolysis of urea	ureD	n/a	Rv1853	n/a	n/a	Mb1884	M. tuberculosis urease operon [Clemens et al., 1995]
	ureG	msmeg_3623	Rv1852	n/a	n/a	Mb1883	
	ureF	msmeg_3624	Rv1851	n/a	n/a	Mb1882	
	ureC	msmeg_3625	Rv1850	n/a	n/a	Mb1881	
	ureB	msmeg_3626	Rv1849	n/a	n/a	Mb1880	
	ureA	msmeg_3627	Rv1848	n/a	n/a	Mb1879	
Unknown	glnA*	msmeg_3827	n/a	n/a	n/a	n/a	E. coli K12 putative glutamine synthetase (b1297)
Unknown	glnA*	msmeg_3828	n/a	n/a	n/a	n/a	[Harth et al., 2005]
Ammonium assimilation	glnA1	msmeg_4290	Rv2220	ML0925	MAP_1962	Mb2244	[Harth et al., 2005]
Post-translational regulation of GSI	glnE	msmeg_4293	Rv2221c	ML1630	MAP_1965c	Mb2245c	[Carroll et al., 2008]
Unknown	glnA2	msmeg_4294	Rv2222c	ML1631	MAP_1966c	Mb2246c	[Harth et al., 2005]
Transcriptional regulation	amtR	msmeg_4300	n/a	n/a	n/a	n/a	amtR C. glutamicum [Jakoby et al., 2000]
Ammonium uptake	amtA	msmeg_4635	n/a	n/a	n/a	n/a	Silicibacter pomeroyi (spo_1578)
Assimilatory nitrate reductase	narI	msmeg_5137	Rv1164	(ML1499)	MAP_2617c	Mb1196	[Khan et al., 2008; Sohaskey, 2008; Sohaskey and Wayne, 2003]
	narJ	msmeg_5138	Rv1163	(ML1500)	MAP_2618c	Mb1195	
	narH	msmeg_5139	Rv1162	(ML1501)	MAP_2619c	Mb1194	
	narG	msmeg_5140	Rv1161	(ML1502)	MAP_2620c	Mb1193	
	narX	n/a	Rv1736c	n/a	n/a	Mb1765c	
Nitrite/nitrate transporter	narK	msmeg_5141	n/a	n/a	n/a	n/a	
	narK1	n/a	Rv2329c	n/a	MAP_2102c	Mb2356c	
	narK2	n/a	Rv1737c	ML0844	n/a	Mb1766c	
	narK3	msmeg_0433	Rv0261c	n/a	MAP_3707c	Mb0267c	
	narU	n/a	Rv0267	n/a	MAP_3712	Mb0273	
Unknown	glnA*	msmeg_5374	n/a	n/a	n/a	n/a	Rhodopseudomonas palustris glnA (RPA0984)

Table 1 (continued)

Enzyme/function	Gene name	M. smegmatis mc² 155	M. tuberculosis H37rv	M. leprae TN	M. avium ssp. paratuberculosis	M. bovis ssp. bovis AF2122/97	Reference/nearest homologs
Glutamate dehydrogenase	gdh	msmeg_5442	n/a	n/a	n/a	n/a	[Börmann et al., 1992]
Ammonium fixation	gltB	msmeg_5594	n/a	n/a	n/a	n/a	C. glutamicum glutamate synthase [Beckers et al., 2001; Schulz et al., 2001]
Transcriptional regulation	glnR	msmeg_5784	Rv0818	(ML2194)	MAP_0649	Mb0841	[Fink et al., 2002; Wray and Fisher, 1993; Wray et al., 1991]
Ammonium transporter	amt1	msmeg_6259	n/a	n/a	n/a	n/a	Pseudomonas syringae (psyr_2277-2273)
Similar to subunits of glutamate synthetase and glutamine synthase; for details, see text	glxA	msmeg_6260	n/a	n/a	n/a	n/a	
	glxB	msmeg_6261	n/a	n/a	n/a	n/a	
	glxC	msmeg_6262	n/a	n/a	n/a	n/a	
	glxD	msmeg_6263	n/a	n/a	n/a	n/a	
Ammonium fixation	gltD	msmeg_6458	n/a	n/a	n/a	n/a	C. glutamicum glutamate synthase [Beckers et al., 2001; Schulz et al., 2001]
	gltB	msmeg_6459	n/a	n/a	n/a	n/a	
Unknown	glnA*	msmeg_6693	n/a	n/a	n/a	n/a	Agrobacterium tumefaciens glnA (Atu4230)

activity measurements in *glnK* deletion strains showed that the ATase GlnE works independently from GlnK in *S. coelicolor* [Hesketh et al., 2002] and *C. glutamicum* [Strösser et al., 2004].

Transcription Control

In parallel to the bioinformatics analysis of genes encoding transporters and enzymes related to nitrogen metabolism, a search was carried out for mycobacterial homologs of known transcriptional regulators of nitrogen control in actinomycetes, namely AmtR of *C. glutamicum* and GlnR of *S. coelicolor* (table 1). Interestingly, both a homolog to the corynebacterial regulator AmtR (42% identity) and a protein with high identity to the regulator of streptomycetes, GlnR (60%) were found in *M. smegmatis*, while in all other examined mycobacterial species, only a homolog for GlnR was found (fig. 1). A phylogenetic tree shows that GlnR homologs of pathogenic mycobacteria form a cluster, while *M. smegmatis* GlnR is more closely related to the corresponding *N. farcinica* and *Rhodococcus* sp. RHA1 protein (fig. 2a). The *M. smegmatis* AmtR is more isolated, only AmtR homologs of corynebacteria are clustered and in summary AmtR proteins are less distributed in actinomycetes (fig. 2b).

When all mycobacterial genomes were screened for known *cis*-acting elements for AmtR and for GlnR, AmtR binding sites could not be identified in any of the mycobacterial genomes. In contrast, using the GlnR motifs of *S. coelicolor*, *Streptomyces avermitilis*, and *Streptomyces*

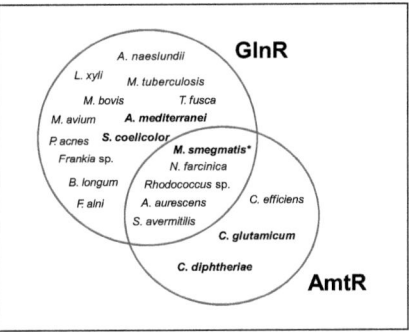

Fig. 1. Distribution of putative nitrogen-dependent transcription regulators GlnR and AmtR in actinomycetes. Bold species names indicate experimental evidence for nitrogen control by the corresponding protein. In *M. smegmatis* (indicated by an asterisk), only the function of GlnR is characterized [Amon et al., 2008], while AmtR function remains unclear [for references of experimentally verified systems see Fink et al., 2002; Jakoby et al., 2000; Nolden et al., 2002; Yu et al., 2006]. Full species names are as follows: *Actinomyces naeslundii, Amycolatopsis mediterranei, Arthrobacter aurescens, Bifidobacterium longum, Corynebacterium diphtheriae, Corynebacterium efficiens, Corynebacterium glutamicum, Frankia alni, Leifsonia xyli, Mycobacterium avium, Mycobacterium bovis, Mycobacterium smegmatis, Mycobacterium tuberculosis, Nocardia farcinica, Propionibacterium acnes, Streptomyces avermitilis, Streptomyces coelicolor,* and *Thermobifida fusca*.

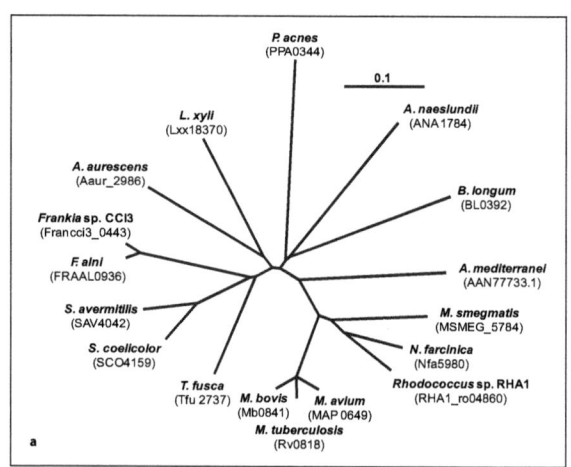

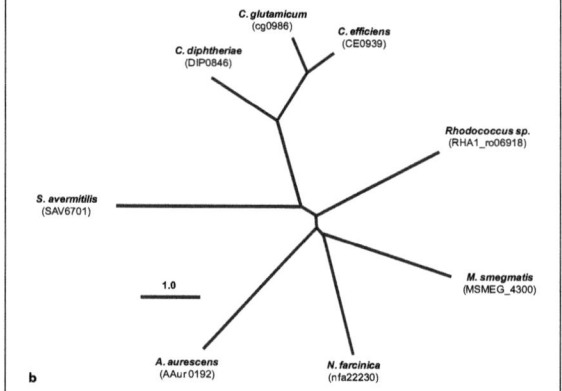

Fig. 2. Phylogenetic trees of the (**a**) GlnR and (**b**) AmtR family of proteins. Unrooted phylogenetic trees were computed with the CLUSTALW software making use of the implemented neighbor joining method with the function for evolutionary distance correction. Evolutionary distances are proportional to the branch length. 17 protein sequences for GlnR and 8 for AmtR homologs were selected as indicated in the figure. For full species names, see figure legend 1.

scabies as query sequences, putative binding sites were detected in the available mycobacterial genomes, including three highly conserved *cis* elements in *M. smegmatis* (fig. 3). These putative binding motifs were located upstream of the *glnA* gene, coding for glutamine synthetase, as well as upstream of *amtB* and *amt1*, ammonium permease-encoding genes.

First experimental evidence supports the idea that GlnR is the crucial regulator of nitrogen control in mycobacteria. As shown by in vitro experiments, purified *S.*

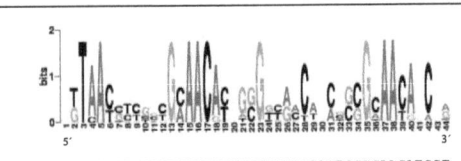

Fig. 3. GlnR-binding motif in mycobacteria. Sequence logo [Crooks et al., 2004] of putative GlnR cis elements identified upstream of the *M. smegmatis* (msmeg), *M. tuberculosis* (H37rv), *M. bovis* (bovis), and *M. avium* (avium) *glnA*, *amtB*, and *amt1* genes (*amt1* is exclusively encoded in the *M. smegmatis* genome).

```
msmeg-glnA   CGTAACGTCGGCGCAACATCGGGTTGACGACTGCGCAACATCGT
H37rv-glnA   AGTAACGTCTGCGCAACACGGGGTTGACTGACGGGCAATATCGG
bovis-glnA   AGTAACGTCTGCGCAACACGGGGTTGACTGACGGGCAATATCGG
avium-glnA   CGTAACGTGCGCGCAACATCGGGTTGACTGACGGGCAACATCTG
msmeg-amtB   GTTCACTTTCCGGAAACGCAACGGCAGCACCGGCGAACGCCGG
H37rv-amtB   GTTAATCCTGATGAAACATGGCGGCACCATCGCCGCAACAACTA
bovis-amtB   GTTAATCCTGATGAAACATGGCGGCACCATCGCCGCAACAACTA
avium-amtB   GTTAATCCCGCCGAAACACAGCGGCACTATCGCCGAAACAACCA
msmeg-amt1   TTTAACCACGCTGCAACACTTGGCGACCATCTCCGTAACAGAAA
```

coelicolor GlnR is able to bind to the *M. tuberculosis glnA* promoter region in electrophoretic mobility shift assays [Tiffert et al., 2008]. Furthermore, deletion mutant analyses showed that transcription of *M. smegmatis amt1, amtB, glnK, glnD* and *glnA* is controlled by GlnR and that the corresponding mutant strain is unable to respond to nitrogen limitation [Amon et al., 2008].

Comparison of the Nitrogen Metabolism-Related Gene Repertoire of Mycobacteria

M. smegmatis is equipped with a variety of genes enabling the uptake and assimilation of nitrogen sources. Compared to the fast-growing *M. smegmatis*, all slow-growing pathogenic members of the genus exhibit a reduced number of genes encoding proteins for nitrogen uptake and assimilation (see table 1 for details and fig. 4 for comparison of *M. smegmatis* and *M. tuberculosis*). This is also due to the fact that *M. smegmatis* seems to have acquired an astonishingly wide range of nitrogen-related genes and gene regions via horizontal gene transfer from a variety of other bacteria, such as *Agrobacterium*, *Burkholderia*, and *Pseudomonas* species. This includes among others a second urease operon, additional ammonium transporters, and a broad variety of glutamine synthetases of various classes and origins (table 1). According to the genomic data, *M. smegmatis* is capable of the active uptake and assimilation of a comparatively wide range of substrates for the extraction of ammonium and further assimilation into central metabolites such as glutamate and glutamine, which is in good concordance to the situation found for the uptake and assimilation of carbohydrates in *M. smegmatis* [Titgemeyer et al., 2007] and thus exhibits a similar repertoire of nitrogen-related genes to that of *C. glutamicum* [Burkovski, 2007; Hänssler and Burkovski, 2008]. Another interesting fact is the co-occurrence of homologs of both regulators of the nitrogen metabolism in actinomycetes in the genome of *M. smegmatis*, namely AmtR and GlnR. While the transcriptional repressor AmtR is the global regulator of nitrogen metabolism in corynebacteria [Walter et al., 2007], this function was only recently shown for the *M. smegmatis* GlnR and its respective target genes *amt1, amtB*, and *glnA* [Amon et al., 2008]; as only homologs of GlnR are found in the genomes of other mycobacteria (fig. 1), the role of AmtR for *M. smegmatis* remains to be further explored.

All mycobacteria investigated exhibit the subunits for a respiratory nitrate reductase (which are nonfunctional pseudogenes in *M. leprae*), while especially the tuberculoid members possess multiple homologs of the *E. coli* nitrite/nitrate antiporters, NarK and NarU. For *M. tuberculosis* it has already been shown that nitrate respiration plays an important role during hypoxia [Sohaskey, 2008], but the additional occurrence of a nitrite reductase, besides its role in detoxification by reduction of nitrite, points to a possible involvement of the enzyme in the complete reduction of nitrate to ammonium and following assimilation, which has been demonstrated for *M. smegmatis* [Khan et al., 2008].

M. leprae unsurprisingly reveals the strongest reduction of nitrogen metabolism-related genes as a process of gene decay termed *'reductive evolution'* [Gómez-Valero

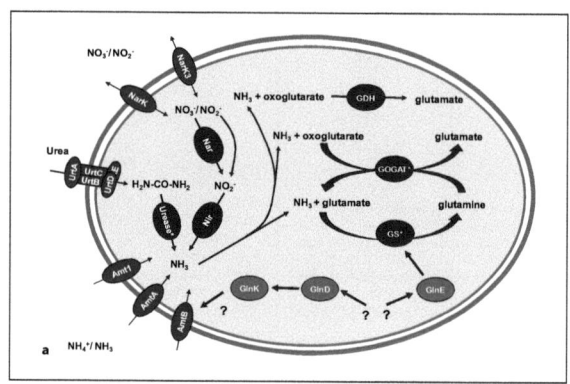

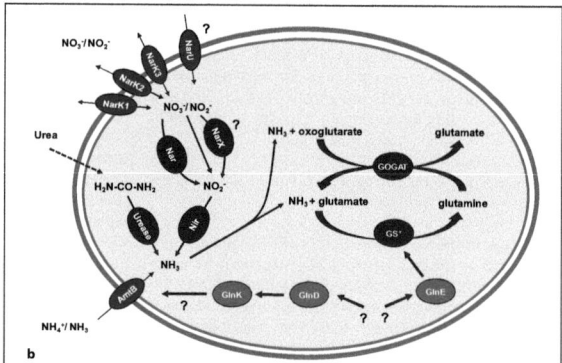

Fig. 4. Comparison of proteins involved in nitrogen uptake, metabolism and control in (**a**) *M. smegmatis* and (**b**) *M. tuberculosis*. Given are the protein names of the respective transporters and enzymes as derived from the in silico analyses (for details, see text). Question marks indicate putative protein-protein interactions and unknown signals. Multiple copies of enzymes are indicated by asterisks.

et al., 2007], resulting in a minimal set of genes required for a functional nitrogen metabolism. This set comprises the genes of the GS/GOGAT pathway *(gltBD, glnA)* as well as the GS ATase *(glnE)*. While we did not find any nitrogen-specific transport systems besides one nitrite/nitrate antiporter (NarK1), the genome of *M. leprae* features putative amino acid and oligopeptide transporters and permeases (data not shown); these substrates could very well represent the main nitrogen sources, taking into account the lifestyle of *M. leprae* as an obligate intracellular parasite [Sassetti et al., 2003; Vissa and Brennan, 2001].

Concluding Remarks

A growing number of genome sequences of different actinomycetes and useful tools and databases for whole genome screening and comparison are already available for research. Due to different studies based on these databases in combination with experimental approaches the transcriptional regulation of nitrogen control in actinomycetes is now well understood. However, there are still a number of open questions regarding the post-translational interactions and modifications between the key players of nitrogen control (fig. 4). For example, the sensory protein for the nitrogen status of the cell that putatively controls the GlnR regulator as well as the signal sensed by this protein is still unknown, and also the function of the P_{II}-type signal transduction proteins is unclear. Furthermore, the existence of two different nitrogen-dependent transcription regulators in the actinomycetes, namely GlnR and AmtR, is still a mystery. The here compiled overview of nitrogen-related genes might thus serve as a blueprint to direct future studies.

Experimental Procedures

In silico Analysis of Mycobacterial Genome Repertoire
A bioinformatic screen of the available genome sequences of *M. smegmatis*, *M. tuberculosis*, *M. leprae*, *M. avium*, and *M. bovis* was performed. For prediction of genes encoding proteins involved in uptake and assimilation of nitrogen sources as well as in nitrogen signal transduction and nitrogen control, genome data obtained from The Institute for Genomic Research (http://www.tigr.org) were screened. To find the most representative mycobacterial homologs of *B. subtilis*, *E. coli*, *C. glutamicum* and *S. coelicolor* proteins, we used single genome protein and nucleotide BLAST available at the following genome server sites: http://genolist.pasteur.fr, TigrBLAST (http://tigrblast.tigr.org/cmr-blast/), CoryneRegNet (http://www.cebitec.uni-bielefeld.de [Baumbach, 2007]) and ScoDB (http://streptomyces.org.uk). Prediction of the possible function was based on the following criteria: (1) protein identity of the mycobacterial protein of more than 30% to the homologous protein of *E. coli* or *B. subtilis*, more than 50% to *S. coelicolor* or *C. glutamicum*, or more than 60% to other mycobacterial proteins, (2) when more than one gene of the corresponding operon was conserved, and (3) most importantly, a solid biochemical analysis of the homologous protein was available. Sequence alignments were conducted with CLUSTALW applying pre-defined algorithms of The European Bioinformatics Institute (EBI) at The European Molecular Biology Laboratory (EMBL) (http://www.ebi.ac.uk/clustalw). Phylogenetic trees were calculated with the neighbour joining method as implemented in CLUSTALW and graphically visualized by importing phylip (*.ph) files into the TREEVIEW software (http://taxonomy.zoology.gla.ac.uk/rod/treeview.hmtl).

Screening for Putative Transcriptional Regulator Binding Sites
All mycobacterial genomes were screened for conserved *cis* elements using the already published binding motifs of the corynebacterial AmtR [Beckers et al., 2005; Walter et al., 2007] and the *Streptomyces* GlnR [Reuther and Wohlleben, 2007] as query sequences using the PreDetector program [Hiard et al., 2007]. The screening for putative *cis* elements was performed by first creating positional weight matrices with different lengths (in bp) for the known *cis* elements of AmtR and GlnR with the Weight Matrix Creation module of PreDetector, then using these weight matrices in the Regulon Prediction module with the genomes of interest. All genomes to be examined were obtained automatically from the NCBI GenBank and downloaded to hard disk for later use.

Acknowledgements

The authors were supported by the Deutsche Forschungsgemeinschaft (SFB473, C12 and D4).

References

Amon J, Bräu T, Grimrath A, Hänßler E, Hasselt K, Höller M, Jeßberger N, Ott L, Szököl J, Titgemeyer F, Burkovski A: Nitrogen control in *Mycobacterium smegmatis*: Nitrogen-dependent expression of ammonium transport and assimilation proteins depends on OmpR-type regulator GlnR. J Bacteriol 2008, Epub ahead of print.

Bange FC, Collins FM, Jacobs WR Jr: Survival of mice infected with *Mycobacterium smegmatis* containing large DNA fragments from *Mycobacterium tuberculosis*. Tuber Lung Dis 1999;79:171–180.

Baumbach J: CoryneRegNet 4.0 – A reference database for corynebacterial gene regulatory networks. BMC Bioinformatics 2007;8:429.

Beckers G, Nolden L, Burkovski A: Glutamate synthase of *Corynebacterium glutamicum* is not essential for glutamate synthesis and is regulated by the nitrogen status. Microbiology 2001;147:2961–2970.

Beckers G, Bendt AK, Krämer R, Burkovski A: Molecular identification of the urea uptake system and transcriptional analysis of urea transporter- and urease-encoding genes in *Corynebacterium glutamicum*. J Bacteriol 2004;186:7645–7652.

Beckers G, Strösser J, Hildebrandt U, Kalinowski J, Farwick M, Krämer R, Burkovski A: Regulation of AmtR-controlled gene expression in *Corynebacterium glutamicum*: mechanism and characterization of the AmtR regulon. Mol Microbiol 2005;58:580–595.

Börmann ER, Eikmanns BJ, Sahm H: Molecular analysis of the *Corynebacterium glutamicum gdh* gene encoding glutamate dehydrogenase. Mol Microbiol 1992;6:317–326.

Burkovski A: I do it my way: Regulation of ammonium uptake and ammonium assimilation in *Corynebacterium glutamicum*. Arch Microbiol 2003a;179:83–88.

Burkovski A: Ammonium assimilation and nitrogen control in *Corynebacterium glutamicum* and its relatives: an example for new regulatory mechanisms in actinomycetes. FEMS Microbiol Rev 2003b:27:617–628.

Burkovski A: Nitrogen metabolism and its regulation; in Bott M, Eggeling L (eds): Handbook of *Corynebacterium glutamicum*. Boca Raton, CRC Press LLC, 2005, pp 333–349.

Burkovski A: Nitrogen control in *Corynebacterium glutamicum*: proteins, mechanisms, signals. J Microbiol Biotechnol 2007;17:187–194.

Carroll P, Pashley CA, Parish T: Functional analysis of GlnE, an essential adenylyl transferase in Mycobacterium tuberculosis. J Bacteriol 2008;DOI:10.1128/JB.00166-08.

Clemens DL, Lee BY, Horwitz MA: Purification, characterization, and genetic analysis of *Mycobacterium tuberculosis* urease, a potentially critical determinant of host-pathogen interaction. J Bacteriol 1995;177:5644–5652.

Cole ST, Brosch R, Parkhill J, Garnier T, Churcher C, Harris D, et al: Deciphering the biology of *Mycobacterium tuberculosis* from the complete genome sequence. Nature 1998; 393:537–544.

Cole ST, Eiglmeier K, Parkhill J, James KD, Thomson NR, Wheeler PR, et al: Massive gene decay in the leprosy bacillus. Nature 2001;409:1007–1011.

Crooks GE, Hon G, Chandonia JM, Brenner SE: WebLogo: a sequence logo generator. Genome Res 2004;14:1188–1190.

Detsch C, Stülke J: Ammonium utilization in *Bacillus subtilis*: transport and regulatory functions of NrgA and NrgB. Microbiology 2003;149:3289–3297.

Dye C, Watt CJ, Bleed DM, Hosseini SM, Raviglione MC: Evolution of tuberculosis control and prospects for reducing tuberculosis incidence, prevalence, and deaths globally. JAMA 2005;293:2767–2775.

Fink D, Weisschuh N, Reuther J, Wohlleben W, Engels A: Two transcriptional regulators GlnR and GlnRII are involved in regulation of nitrogen metabolism in *Streptomyces coelicolor* A3(2). Mol Microbiol 2002;46:331–347.

Garnier T, Eiglmeier K, Camus JC, Medina N, Mansoor H, Pryor M, et al: The complete genome sequence of *Mycobacterium bovis*. Proc Natl Acad Sci USA 2003;100:7877–7882.

Gómez-Valero L, Rocha EPC, Latorre A, Silva FJ: Reconstructing the ancestor of *Mycobacterium leprae*: The dynamics of gene loss and genome reduction. Genome Res 2007;17: 1178–1185.

Hänßler E, Burkovski A: Molecular mechanisms of nitrogen control in corynebacteria; in Burkovski A (ed): Corynebacteria: Genomics and Molecular Biology. Wymondham, Caister Academic Press, 2008, pp 183–201.

Harth G, Clemens DL, Horwitz AA: Glutamine synthetase of *Mycobacterium tuberculosis*: extracellular release and characterization of its enzymatic activity. Proc Natl Acad Sci USA 1994;91:9342–9346.

Harth G, Maslesa-Galic S, Tullius MV, Horwitz AA: All four *Mycobacterium tuberculosis* glnA genes encode glutamine synthetase activities but only GlnA1 is abundantly expressed and essential for bacterial homeostasis. Mol Microbiol 2005;58:1157–1172.

Hesketh A, Fink D, Gust B, Rexer H-U, Scheel B, Chater K, Wohlleben W, Engels A: The GlnD and GlnK homologues of *Streptomyces coelicolor* A3(2) are functionally dissimilar to their nitrogen regulatory system counterparts from enteric bacteria. Mol Microbiol 2002;46:319–330.

Hiard S, Maree R, Colson S, Hoskisson PA, Titgemeyer F, van Wezel G, Joris B, Wehenkel L, Rigali S: PREDetector: a new tool to identify regulatory elements in bacterial genomes. Biochem Biophys Res Commun 2007;357:861–864.

Hotter GS, Mouat P, Collins DM: Independent transcription of glutamine synthetase (glnA2) and glutamine synthetase adenylyltransferase (glnE) in *Mycobacterium bovis* and *Mycobacterium tuberculosis*. Tuberculosis DOI:10.1016/j.tube.2008.02.006.

Jakoby M, Krämer R, Burkovski A: Nitrogen regulation in *Corynebacterium glutamicum*: Isolation of genes involved and biochemical characterization of the corresponding proteins. FEMS Microbiol Lett 1999;173:303–310.

Jakoby M, Nolden L, Meier-Wagner J, Krämer R, Burkovski A: AmtR, a global repressor in the nitrogen regulation system of *Corynebacterium glutamicum*. Mol Microbiol 2000;37: 964–977.

Khan A, Akhtar S, Ahmad J, Sarkar D: Presence of a functional nitrate pathway in *Mycobacterium smegmatis*. Microb Pathog 2008;44: 71–77.

Li L, Bannantine JP, Zhang Q, Amonsin A, Max BJ, Alt D, Banerji N, Kanjilal S, Kapur V: The complete genome sequence of *Mycobacterium avium* subspecies *paratuberculosis*. Proc Natl Acad Sci USA 2005;102:12344–12349.

Meier-Wagner J, Nolden L, Jakoby M, Siewe R, Krämer R, Burkovski A: Multiplicity of ammonium uptake systems in *Corynebacterium glutamicum*: role of Amt and AmtB. Microbiology 2001;147:135–143.

Nolden L, Beckers G, Burkovski A: Nitrogen assimilation in *Corynebacterium diphtheriae*: pathways and regulatory cascades. FEMS Microbiol Lett 2002;208:287–293.

Nordqvist A, Nilsson MT, Röttger S, Odell LR, Krajewski WW, Andersson CE, Larhed M, Mowbray SL, Karlén A: Evaluation of the amino acid binding site of *Mycobacterium tuberculosis* glutamine synthetase for drug discovery. Bioorg Med Chem 2008;6:5501–5513.

Parish T, Stoker NG: glnE is an essential gene in *Mycobacterium tuberculosis*. J Bacteriol 2000;182:5715–5720.

Pashley CA, Brown AC, Robertson D, Parish T: Identification of the *Mycobacterium tuberculosis* glnE promoter and its response to nitrogen availability. Microbiology 2006;152: 2727–2734.

Pelicic V, Reyrat JM, Gicquel B: Genetic advances for studying *Mycobacterium tuberculosis* pathogenicity. Mol Microbiol 1998;28:413–420.

Read R, Pashley CA, Smith D, Parish T: The role of GlnD in ammonium assimilation in *Mycobacterium tuberculosis*. Tuberculosis 2007; 87:384–390.

Reuther J, Wohlleben W: Nitrogen metabolism in *Streptomyces coelicolor*: Transcriptional and post-translational regulation. J Mol Microbiol Biotechnol 2007;12:139–146.

Rexer HU, Schäberle T, Wohlleben W, Engels A: Investigation of the functional properties and regulation of three glutamine synthetase-like genes in *Streptomyces coelicolor* A3(2). Arch Microbiol 2006;186:447–458.

Sassetti CM, Boyd DH, Rubin EJ: Genes required for mycobacterial growth defined by high density mutagenesis. Mol Microbiol 2003;48: 77–84.

Schulz AA, Collett HJ, Reid SJ: Regulation of glutamine synthetase and glutamate synthase in *Corynebacterium glutamicum* ATCC 13032. FEMS Microbiol Lett 2001;205:361–367.

Sohaskey CD: Nitrate enhances the survival of *Mycobacterium tuberculosis* during inhibition of respiration. J Bacteriol 2008;190: 2981–2986.

Sohaskey CD, Wayne LG: Role of narK2X and narGHJI in hypoxic upregulation of nitrate reduction by *Mycobacterium tuberculosis*. J Bacteriol 2003;185:7247–7256.

Strösser J, Lüdke A, Schaffer S, Krämer R, Burkovski A: Regulation of GlnK activity: modification, membrane sequestration, and proteolysis as regulatory principles in the network of nitrogen control in *Corynebacterium glutamicum*. Mol Microbiol 2004;54: 132–147.

Tiffert Y, Supra P, Wurm R, Wohlleben W, Wagner R, Reuther J: The *Streptomyces coelicolor* GlnR regulon: identification of new GlnR targets and evidence for a central role of GlnR in nitrogen metabolism in actinomycetes. Mol Microbiol 2008;67:436–446.

Titgemeyer F, Parche S, Mahfoud M, Amon J, Bail J, Schlicht M, Hillmann D, Rehm N, Stephan J, Walter B, Burkovski A, Niederweis M: Sugar transport in *Mycobacterium smegmatis* and *Mycobacterium tuberculosis*. J Bacteriol 2007;189:5903–5915.

Tullius MV, Harth G, Horwitz AA: Glutamine synthetase GlnA1 is essential for growth of *Mycobacterium tuberculosis* in human THP-1 macrophages and guinea pigs. Infect Immun 2003;71:3927–3936.

Vissa VD, Brennan PJ: The genome of *Mycobacterium leprae*: a minimal mycobacterial genome set. Genome Biol 2001;2:1023.1–1023.8.

Walter B, Hänßler E, Kalinowski J, Burkovski A: Nitrogen metabolism and nitrogen control in corynebacteria: variations of a common theme. J Mol Microbiol Biotechnol 2007;12: 131–138.

Walter B, Küspert M, Ansorge A, Krämer R, Burkovski A: Dissection of ammonium uptake systems in *Corynebacterium glutamicum*: mechanism and energetics of AmtA and AmtB. J Bacteriol 2008;190:2611–2614.

Wang H, Gunsalus RP: The nrfA and nirB nitrite reductase operons in *Escherichia coli* are expressed differently in response than to nitrate to nitrite. J Bacteriol 2000;182:5813–5822.

Wray LV, Atkinson MR, Fisher SH: Identification and cloning of the glnR locus, which is required for transcription of the glnA gene in *Streptomyces coelicolor* A3(2). J Bacteriol 1991;173:7351–7360.

Wray LV, Fisher SH: The *Streptomyces coelicolor* glnR gene encodes a protein similar to other bacterial response regulators. Gene 1993; 130:145–150.

Yu H, Peng W, Liu Y, Wu T, Yao Y, Cui M, Jiang W, Zhao G-P: Identification and characterization of glnA promoter and its corresponding trans-regulating protein GlnR in the rifamycin SV producing actinomycete, *Amycolatopsis mediterranei* U32. Acta Biochim Biophys Sin 2006;38:831–843.

Nitrogen Control in *Mycobacterium smegmatis*: Nitrogen-Dependent Expression of Ammonium Transport and Assimilation Proteins Depends on the OmpR-Type Regulator GlnR[▽][†]

Johannes Amon, Tanja Bräu, Aletta Grimrath, Eva Hänßler, Kristin Hasselt, Martina Höller, Nadja Jeßberger, Lisa Ott, Juraj Szököl, Fritz Titgemeyer, and Andreas Burkovski*

Lehrstuhl für Mikrobiologie, Friedrich-Alexander-Universität Erlangen-Nürnberg, Erlangen, Germany

Received 23 June 2008/Accepted 28 July 2008

The effect of nitrogen regulation on the level of transcriptional control has been investigated in a variety of bacteria, such as *Bacillus subtilis*, *Corynebacterium glutamicum*, *Escherichia coli*, and *Streptomyces coelicolor*; however, until now there have been no data for mycobacteria. In this study, we found that the OmpR-type regulator protein GlnR controls nitrogen-dependent transcription regulation in *Mycobacterium smegmatis*. Based on RNA hybridization experiments with a wild-type strain and a corresponding mutant strain, real-time reverse transcription-PCR analyses, and DNA binding studies using cell extract and purified protein, the *glnA* (msmeg_4290) gene, which codes for glutamine synthetase, and the *amtB* (msmeg_2425) and *amtI* (msmeg_6259) genes, which encode ammonium permeases, are controlled by GlnR. Furthermore, since *glnK* (msmeg_2426), encoding a PII-type signal transduction protein, and *glnD* (msmeg_2427), coding for a putative uridylyltransferase, are in an operon together with *amtB*, these genes are part of the GlnR regulon as well. The GlnR protein binds specifically to the corresponding promoter sequences and functions as an activator of transcription when cells are subjected to nitrogen starvation.

Almost all of the macromolecules in a bacterial cell, including proteins, nucleic acids, and cell wall components, contain nitrogen. Consequently, prokaryotes have developed elaborate mechanisms to provide an optimal nitrogen supply for metabolism and to overcome and survive under nitrogen starvation conditions. We are especially interested in nitrogen metabolism and regulation in mycolic acid-containing actinomycetes. In the actinomycete group, the uptake and assimilation of nitrogen sources have been studied most intensively for *Corynebacterium glutamicum* (for reviews, see references 4, 5, 6, 9) and *Streptomyces coelicolor* (for a review, see reference 21).

In corynebacteria, like the amino acid-producing strains of *C. glutamicum* and *Corynebacterium efficiens* or the pathogen *Corynebacterium diphtheriae*, the expression of genes coding for proteins involved in uptake and assimilation of nitrogen sources is under the control of a central regulatory protein, the TetR-type regulator AmtR (14, 17, 28). In *C. glutamicum* AmtR blocks transcription of at least 33 genes (2), including genes which encode ammonium transporters (*amtA* and *amtB*), ammonium assimilation enzymes (*glnA* and *gltBD*), creatinine (*crnT* and *codA*) and urea (*urtABCDE* and *ureABCEFGD*) transport and metabolism, a number of biochemically uncharacterized enzymes and transport systems, and signal transduction proteins (*glnD* and *glnK*).

For *S. coelicolor*, it has been shown that various genes encoding nitrogen metabolism-related proteins are under the control of a different regulatory protein, the OmpR-type regulator GlnR (8, 26). GlnR acts as a transcriptional activator for at least 15 genes encoding proteins related to nitrogen uptake, metabolism, and regulation, as well as proteins with unknown functions. The GlnR-regulated genes include *amtB* encoding a putative ammonium uptake system, *glnK* and *glnD* coding for signal transduction proteins, the glutamine synthetase-encoding *glnA* and *glnII* genes, *ureA* encoding urease subunit gamma, and *nirB* coding for a large subunit of a putative nitrite reductase. Interestingly, a second GlnR regulator is encoded in the *S. coelicolor* genome; this regulator, GlnRII, also binds to the upstream regions of *glnA*, *glnII*, *amtB*, *glnK*, and *glnD* (8). The regulatory function of this protein is still unknown.

Less functional information is available for mycobacteria (10). Previous work on nitrogen uptake and assimilation concentrated mainly on the major pathogenic member of the genus, *Mycobacterium tuberculosis*. While extensive data are available for glutamine synthetases, especially the physiologically crucial GlnA1 enzyme (12), the associated adenylyltransferase GlnE (7, 18), and the uridylyl transferase GlnD (20), not much information is available for the transcriptional regulation of nitrogen metabolism in mycobacteria (19). In *Mycobacterium smegmatis*, homologs of all previously characterized genes encoding glutamine synthetases in *M. tuberculosis* are present in the genome (GlnA1, MSMEG_4290; GlnA2, MSMEG_4294; GlnA3, MSMEG_3561; GlnA4, MSMEG_2595), as are open reading frames encoding additional glutamine synthetase-like proteins with unknown physiological functions (e.g., MSMEG_1116, MSMEG_3827, MSMEG_5374, and MSMEG_6693).

In order to investigate the control of ammonium assimilation in *M. smegmatis*, we performed an in silico analysis using the previously described genome sequence of *M. smegmatis* mc²155 encoding homologs of known nitrogen control proteins

* Corresponding author. Mailing address: Lehrstuhl für Mikrobiologie, Friedrich-Alexander-Universität Erlangen-Nürnberg, Staudtstr. 5, 91058 Erlangen, Germany. Phone: 49 9131 85 28086. Fax: 49 9131 85 28082. E-mail: aburkov@biologie.uni-erlangen.de.
† Supplemental material for this article may be found at http://jb.asm.org/.
▽ Published ahead of print on 8 August 2008.

and their corresponding *cis*-acting elements. The results of this analysis were verified experimentally.

MATERIALS AND METHODS

Bioinformatic analyses. To screen genes encoding well-investigated nitrogen transcriptional regulators (*amtR* of *C. glutamicum*, as well as *glnR* and *glnRII* of *S. coelicolor*), genome data were obtained from The Institute for Genomic Research (http://www.tigr.org). To find the most representative homologs, we used single-genome protein BLAST available at the following genome server sites for well-characterized bacterial species: TigrBLAST (http://tigrblast.tigr.org/cmr-blast/) for mycobacterial species, CoryneRegNet (http://www.cebitec.uni-bielefeld.de) [1] for corynebacteria, and the *S. coelicolor* database ScoDB (http://streptomyces.org.uk). Sequence alignment was performed with CLUSTALW by using predefined algorithms of The European Bioinformatics Institute at The European Molecular Biology Laboratory (www.ebi.ac.uk/clustalw).

In a parallel analysis, all mycobacterial genomes available at the NCBI GenBank were screened for conserved *cis* elements using the previously described binding motifs of corynebacterial AmtR [2, 28] and *Streptomyces* GlnR [21] as query sequences. The PreDetector program [13] was used to screen the available mycobacterial genomes after successful positive control analyses were performed with the AmtR and GlnR binding motifs in the genomes of *C. glutamicum* and *S. coelicolor*, respectively. Screening for putative *cis* elements was performed by first creating positional weight matrices with different lengths (in bp) for the known *cis* elements of AmtR and GlnR with the Weight Matrix Creation module of PreDetector and then using these weight matrices in the Regulon Prediction module with the genomes of interest.

Strains and growth conditions. Bacteria were routinely grown at 37°C in baffled flasks with agitation. Mycobacterial strains were grown in Middlebrook 7H9 liquid medium (Difco Laboratories) (containing [per 900 ml] approximately 0.5 g ammonium sulfate, 0.5 g L-glutamic acid, 0.1 g sodium citrate, 1.0 mg pyridoxine, 0.5 mg biotin, 2.5 g disodium phosphate, 1.0 g monopotassium phosphate, 0.04 g ferric ammonium citrate, 0.05 g magnesium sulfate, 0.5 mg calcium chloride, 1.0 mg zinc sulfate, and 1.0 mg copper sulfate) supplemented with 0.2% glycerol and 0.05% Tween 80 or on Middlebrook 7H9 medium (Difco Laboratories) containing 1.5% agar supplemented with 0.2% glycerol. When appropriate, antibiotics were added at the following concentrations: hygromycin, 100 μg ml^{-1} for *Escherichia coli* and 50 μg ml^{-1} for *M. smegmatis*; kanamycin, 30 μg ml^{-1} for *E. coli* and 10 μg ml^{-1} for *M. smegmatis*; and streptomycin, 400 μg ml^{-1} for *M. smegmatis*. In order to study the effects of nitrogen starvation, a fresh *M. smegmatis* culture was used to inoculate Middlebrook 7H9 medium for overnight growth. This culture, which had an optical density at 600 nm (OD$_{600}$) of approximately 2 to 3 after overnight growth, was used to inoculate fresh Middlebrook 7H9 medium to obtain an OD$_{600}$ of approximately 0.2, and cells were grown for 10 to 11 h until the exponential growth phase was reached (OD$_{600}$, approximately 0.6 to 0.8). To induce nitrogen starvation, cells were harvested by centrifugation, washed, and resuspended in prewarmed Middlebrook 7H9 medium without a nitrogen source (Middlebrook 7H9 medium lacking ammonium sulfate and glutamic acid and containing iron citrate instead of iron ammonium citrate). As a control, unstarved cells were harvested, washed, and transferred using Middlebrook 7H9 medium. Alternatively, nitrogen limitation was induced by addition of methionine sulfoximine (MSX) (final concentration, 200 μM) to Middlebrook 7H9 medium. This substrate analog inhibits glutamine synthetase activity and consequently blocks nitrogen metabolism [15, 16].

General molecular biology techniques. For plasmid isolation, transformation, and cloning, standard techniques were used [22]. *E. coli* strain JM109 [31] was used as the cloning host. Plasmids were subsequently transferred into competent *M. smegmatis* cells by electroporation. Chromosomal DNA was extracted from 100-ml cultures grown to stationary phase as described previously [3]. DNA sequence analyses were carried out using a BigDye Terminator V3.1 cycle sequencing kit (Perkin Elmer).

Construction of a *glnR* deletion strain. In order to generate a *glnR* deletion, two DNA fragments flanking the *glnR* gene (1 kb up- and downstream of *glnR*) were PCR amplified using chromosomal DNA from *M. smegmatis* strain SMR5 as the template. For subsequent cloning in the pML814 vector (ColE1 origin, *FRT-hyg-FRT rpsL*, Amp^r Hyg^r; 6,220 bp; general deletion vector; kindly provided by M. Niedermeiss), SwaI and PacI restriction sites were introduced into the primer sequences used for amplification of the upstream sequence of *glnR* and SpeI and PmeI restriction sites were introduced into the primer sequences used for amplification of the downstream sequence of *glnR* (see Table S1 in the supplemental material). The resulting plasmid, pML814Δ*glnR*, carried an *FRT-hyg-FRT* expression cassette [24] flanked by the sequences upstream and downstream of *glnR*. *M. smegmatis* SMR5, a streptomycin-resistant derivative of *M. smegmatis* mc^{2}155, was transformed with plasmid pML814Δ*glnR*, and transformants were selected on hygromycin-containing plates to obtain a single crossover [24]. After verification of the single-crossover event using PCR, cells were selected on plates containing hygromycin and streptomycin. Clones on these plates should have lost the vector and should have had the *FRT-hyg-FRT* cassette integrated into the chromosome. After verification of the double-crossover event using PCR, the FLP recombinase was used to specifically remove the *hyg* gene from the chromosome, generating a marker-free deletion mutant and allowing the *hyg* gene to be reused as resistance marker. Selection of clones was carried out using plates containing hygromycin and streptomycin. Deletion of *glnR* in the resulting strain, MH1, was verified by PCR and Southern blotting (data not shown).

Construction of antisense probes. For generation of antisense probes, internal DNA fragments of the corresponding genes were amplified by PCR (the primers used for the different probes are shown in Table S1 in the supplemental material). The reverse primer encoded the promoter region for T7 polymerase, which allowed in vitro transcription of probes using T7 polymerase.

RNA preparation, hybridization analyses, and real-time reverse transcription-PCR (RT-PCR). *M. smegmatis* RNA was prepared from 6-ml culture samples using a NucleoSpin RNA II kit (Macherey Nagel, Düren, Germany). If necessary, a second DNase digestion was performed with Turbo DNase (Ambion) to completely remove the chromosomal DNA. RNA samples were stored at −80°C.

Antisense probes that were between 0.2 and 0.5 kb long and were used for analysis of gene transcription were generated by PCR and subsequent labeling with a digoxigenin (DIG) RNA-labeling mixture (Roche, Mannheim, Germany) and T7 polymerase (NEB, Frankfurt, Germany). RNA (1 μg per time point) was spotted onto nylon membranes using a Schleicher & Schuell (Dassel, Germany) Minifold I dot blotter. Hybridization of DIG-labeled RNA probes was detected with X-ray film (Amersham Hyperfilm MP; GE Healthcare) using alkaline phosphatase-conjugated anti-DIG Fab fragments and chloro-5-substituted adamantyl-1,2-dioxetane phosphate (CSPD) as a light-emitting substrate as recommended by the supplier (Roche, Mannheim, Germany). All experiments were carried out at least in duplicate with independent cultures (biological replicates).

For real-time RT-PCR, an MyiQ single-color real-time PCR detection system (Bio-Rad, Munich, Germany), a QuantiTect SYBR green RT-PCR kit (Qiagen, Hilden, Germany), primers at a concentration of 1 μM, and 100 ng of template RNA were used. RT was carried out at 50°C for 30 min, and the reverse transcriptase was inactivated and the polymerase was activated by 15 min of incubation at 95°C. The PCR was carried out by using 40 cycles of DNA denaturation for 15 s at 94°C, primer annealing for 30 s at 60°C, and DNA polymerization for 15 s at 72°C. The PCR was followed by a melting curve program (55 to 100°C with a heating rate of 1°C per 10 s) and then a cooling program (25°C). No-template controls were included with all reactions. Data were analyzed using the MyiQ single-color real-time PCR detection system software.

Construction of *glnR*-carrying plasmids. For complementation assays, the *glnR* gene was amplified by PCR using chromosomal DNA of *M. smegmatis* as the template and oligonucleotides glnR-fw-PacI and glnR-rev-SwaI (see Table S1 in the supplemental material). The PCR products were ligated to plasmid pMN016 (psmyc-*mspA*, ColE1 origin, PAL5000 origin, Hyg^r; 6,164 bp), [25] using the SwaI and PacI sites, which was introduced by using the oligonucleotide primers. The resulting plasmid, pMN016-*glnR*, was sequenced and used as a control.

For purification of GlnR for gel retardation assays, the *glnR* gene was amplified by PCR (for the primers used, see Table S1 in the supplemental material) and ligated to plasmid pQE70 (Qiagen, Hilden, Germany) in order to add a C-terminal His$_6$ tag to the expressed protein, resulting in plasmid pQE-*glnR*-His. For expression of His-tagged GlnR in *M. smegmatis*, the corresponding DNA fragment was excised from plasmid pQE-*glnR*-His by HindIII/SphI restriction and ligated to dephosphorylated and HindIII/SphI-cut plasmid pMN016, resulting in plasmid pMN016-*glnR*-His.

Purification of GlnR and gel retardation experiments. For gel shift assays, protein extracts were prepared from *M. smegmatis* wild-type strain SMR5 carrying plasmid pMN016-*glnR*-His. Cells were cultivated in Middlebrook 7H9 medium containing antibiotics as described above, harvested by centrifugation (4,000 × *g*, 15 min, 4°C), and suspended in 300 mM NaCl–50 mM NaH$_2$PO$_4$–20 mM imidazole (pH [NaOH] 8.0) (2 ml g cells^{-1}) containing lysozyme (2 mg ml^{-1}) and Complete protease inhibitor as recommended by the supplier (Roche, Mannheim, Germany). Cells were subsequently disrupted by ultrasonic treatment. Cell debris was removed by centrifugation (14,000 × *g*, 30 min, 4°C), and the protein extract was loaded on a 5-ml Ni-nitrilotriacetic acid column in a chromatography apparatus (Äkta prime; GE Healthcare, Munich, Germany). Washing and subsequent elution with 300 mM NaCl–50 mM NaH$_2$PO$_4$–500 mM imidazole (pH [NaOH] 8.0) were carried out as recommended by the supplier of the Ni-nitrilotriacetic acid matrix (GE Healthcare, Munich, Germany); the purified protein was stored at −80°C.

Target DNA for the gel shift assays was synthesized by PCR (for the primers used, see Table S1 in the supplemental material) and was purified by agarose gel electrophoresis. To label the DNA and to prepare the reaction mixture for the gel shift assay, a DIG gel shift kit (Roche, Mannheim) was used as recommended by the supplier. Separation by gel electrophoresis was performed in native 6% polyacrylamide gels (Anamed Electrophorese GmbH, Darmstadt, Germany) using 0.5× Tris-borate-EDTA buffer as the running buffer. Subsequently, the labeled DNA was transferred to a nylon membrane (Roche, Mannheim, Germany) by electroblotting as described in the protocol of the DIG gel shift kit (Roche, Mannheim, Germany). For detection of the labeled DNA, X-ray film was used.

DNA affinity purification of GlnR. For DNA affinity purification with magnetic beads, target DNAs were amplified by PCR using biotinylated oligonucleotides (see Table S1 in the supplemental material). For preparation of DNA-coated magnetic beads, M-280 Dynabeads coated with streptavidin (Dynal, Oslo, Norway) were washed and resuspended in 200 μl of 1× DNA binding buffer (5 mM Tris-HCl [pH 7.5], 0.5 mM EDTA, 1 M NaCl), and biotinylated target DNA was added. After incubation for 30 min at room temperature to allow DNA binding to the streptavidin-coated magnetic beads, the beads were washed three times with 500 μl of 1× DNA binding buffer and subsequently stored at 4°C in Tris-EDTA buffer containing 0.02% NaN$_3$. Before use, the DNA-coated magnetic beads were washed three times in 1 ml ice-cold protein binding buffer (phosphate-buffered saline) (pH 7.4) (22). For preparation of protein, *M. smegmatis* wild-type strain SMR5 was grown in 500 ml Middlebrook 7H9 medium until the OD$_{600}$ was approximately 2. Cells were harvested by centrifugation (4,000 × g, 15 min, 4°C) and resuspended in 4 ml ice-cold lysis buffer (50 mM Tris-HCl, 70 mM KCl, 1 mM EDTA, 1 mM dithiothreitol, 10% glycerol, 400 μl Roche Complete protease inhibitor; pH 8.0). The cell suspension was transferred to tubes containing glass beads, and the cells were disrupted by vigorous shaking at 6.5 m s^{-1} for 1 min using a FastPrep FP120 instrument (Q-BIOgene, Heidelberg, Germany). The cell debris and glass beads were removed by centrifugation (14,000 × g, 4 min, 4°C) before the membranes were separated by ultracentrifugation (267,000 × g, 30 min, 4°C). One-milliliter aliquots of the supernatant containing the cytoplasmic proteins were incubated with the DNA-coated beads for 45 min at 4°C with continuous agitation. The beads were separated from the protein extract and washed twice with 500 μl ice-cold protein binding buffer. This procedure was repeated three times. Finally, the proteins bound to the DNA were eluted. To do this, the beads were resuspended in 20-μl portions of elution buffer (phosphate-buffered saline with the sodium chloride concentration increasing from 200 mM to 1,000 mM in 100 mM steps). The elution fractions were analyzed by sodium dodecyl sulfate-polyacrylamide gel electrophoresis (SDS-PAGE) (23), and proteins were identified by peptide mass fingerprinting.

Mass spectrometry. Matrix-assisted laser desorption ionization–time of flight mass spectrometry (MALDI-TOF MS) and peptide mass fingerprint analysis were carried out by the bioanalytics service unit at the Center for Molecular Medicine of Cologne.

RESULTS

In silico analysis of transcription regulators and corresponding binding sites. A bioinformatic analysis was carried out to identify *M. smegmatis* homologs of known global transcriptional regulators of nitrogen control in actinomycetes. Interestingly, both a homolog of the *C. glutamicum* regulator AmtR and a protein with a high level of identity to GlnR, the regulator of nitrogen metabolism in *S. coelicolor*, were found in *M. smegmatis*. The *M. smegmatis* AmtR shows 42% identity to *C. glutamicum* AmtR, and GlnR exhibits 55% amino acid identity to *S. coelicolor* GlnR. However, when the previously described genome sequences of *M. tuberculosis*, *Mycobacterium bovis*, and *Mycobacterium avium* were screened, only a homolog of *glnR* was found. Obviously, GlnR proteins are highly conserved in actinobacteria (Fig. 1) (26), and based on their wide distribution in different genera, including *Actinomyces*, *Amycolatopsis*, *Arthrobacter*, *Bifidobacterium*, *Frankia*, *Leifsonia*, *Mycobacterium*, *Nocardia*, *Rhodococcus*, *Streptomyces*, and *Thermobifida*, they are most likely involved in nitrogen regulation in mycobacteria. While no homologs of the *S. coelicolor* GlnR protein have been found in the available corynebacterial genomes, the organization of the genome sequences of the genera mentioned above is conserved and the genomes contain an apparently monocistronic operon with no open reading frames close to the corresponding *glnR* genes.

Using a parallel approach, in which the genomes of *M. smegmatis*, *M. tuberculosis*, *M. bovis*, and *M. avium* were screened for known *cis*-acting elements of AmtR (2, 28), corresponding binding sites could not be identified. In contrast, using the GlnR motifs of *S. coelicolor*, *Streptomyces avermitilis*, and *Streptomyces scabies* (21) as query sequences, putative binding sites were detected in the available mycobacterial genomes, and these sites included three highly conserved *cis* elements in *M. smegmatis* (Fig. 2). These putative binding motifs were located upstream of the *glnA* gene (msmeg_4290), which codes for glutamine synthetase and is homologous to the *M. tuberculosis glnA1* gene coding for the physiologically crucial GS enzyme, as well as upstream of *amtB* (msmeg_2425) and *amt1* (msmeg_6259), which are ammonium permease-encoding genes. No GlnR *cis* elements were detected upstream of urease subunit-encoding genes (msmeg_1091 and msmeg_3627), glutamate synthase-encoding genes (msmeg_3225, msmeg_5594, and msmeg_6459), or other genes encoding putative glutamine synthetases (msmeg_1116, msmeg_2595, msmeg_3561, msmeg_3827, msmeg_4294, msmeg_5374, and msmeg_6693).

The *M. smegmatis amtB* gene (msmeg_2425) is localized in a gene cluster together with *glnK* (msmeg_2426), which encodes a PII-type signal transduction protein, and *glnD* (msmeg_2427), which encodes a putative uridylyltransferase. By using RT-PCR, a common transcript of these three genes was detected (data not shown), adding *glnK* and *glnD* to the putative GlnR regulon.

DNA affinity purification of GlnR. As described above, putative GlnR binding motifs were detected upstream of the *glnA* gene, as well as upstream of *amtB* and *amt1*. The corresponding promoter regions, including the GlnR binding site, were amplified by PCR and bound to magnetic beads. The beads were incubated with *M. smegmatis* cytoplasmic proteins. Subsequently, bound proteins were eluted, separated by SDS-PAGE (Fig. 3), and identified by MALDI-TOF MS and peptide mass fingerprinting (Table 1). In addition to transcription termination factor Rho, DNA topoisomerase I, and a protein annotated as a hypothetical protein, GlnR was isolated using this approach. In fact, GlnR was purified with all promoter DNA sequences used, supporting the hypothesis that there is transcriptional control of these genes by this regulator. Although binding of GlnR to the promoter regions used seemed to be specific, as indicated by the high NaCl concentrations necessary for elution, in principle, binding of GlnR could also be explained by a general DNA binding property of the protein. Therefore, genetic analyses were carried out to further characterize the GlnR function.

GlnR-dependent transcriptional response to nitrogen starvation. To validate the data obtained by in silico analyses and DNA affinity purification and to elucidate the putative role of GlnR in nitrogen control in *M. smegmatis*, a *glnR* deletion was introduced into the *M. smegmatis* genome. During growth on

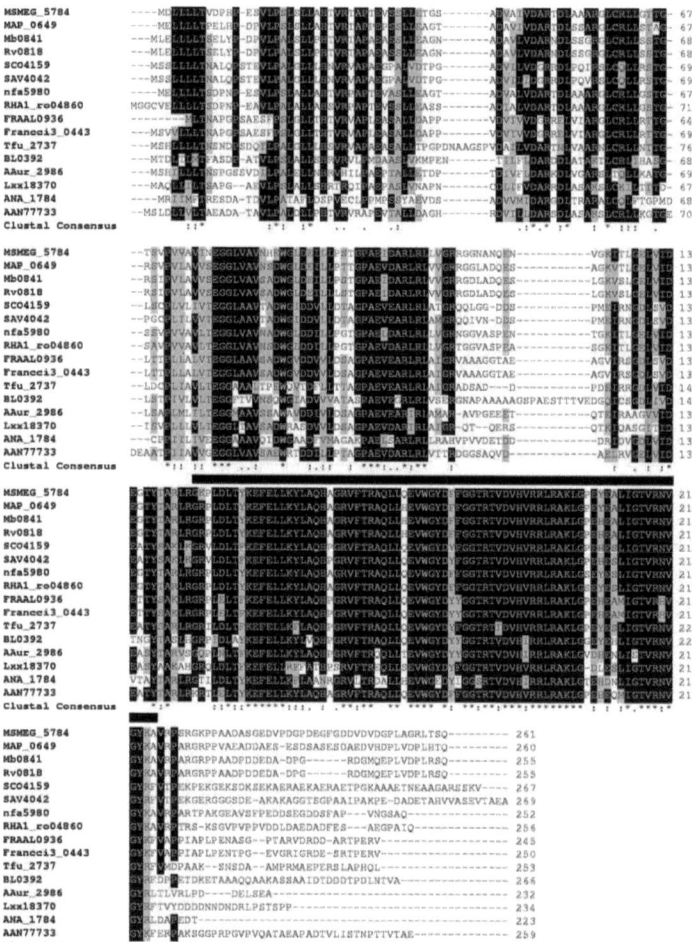

FIG. 1. Alignment of GlnR proteins from various actinomycetes. Abbreviations used for GlnR homologs with annotation numbers: MSMEG, *Mycobacterium smegmatis*; MAP, *Mycobacterium avium* subsp. *paratuberculosis*; Mb, *Mycobacterium bovis*; Rv, *Mycobacterium tuberculosis* H37rv; SCO, *Streptomyces coelicolor*; SAV, *Streptomyces avermitilis*; nfa, *Nocardia farcinica*; RHA1, *Rhodococcus* sp. strain RHA1; FRAAL, *Frankia alni* ACN14a; Francci3, *Frankia* sp. strain Cci3; Tfu, *Thermobifida fusca*; BL, *Bifidobacterium longum*; AAur, *Arthrobacter aurescens*; Lxx, *Leifsonia xylii*; ANA, *Actinomyces naeslundii*; AAN77733, *Amycolatopsis mediterranei*. Identical amino acid residues are indicated by a black background, and similar amino acids are indicated by a gray background. *M. smegmatis* GlnR contains a putative phosphorylation pocket, located at amino acid residues 43 to 53, and a C-terminal DNA binding domain (helix turn helix), located at amino acid residues 144 to 218 (indicated by a bar above the sequence). *, identical residues; :, conserved substitutions/residues; ., semiconserved substitutions.

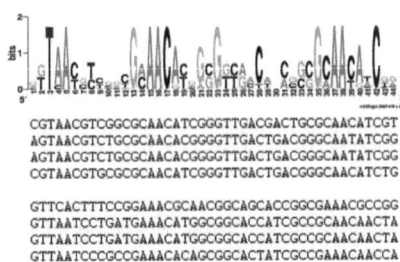

```
msmeg-glnA    CGTAACGTCGGCGCAACATCGGGTTGACGACTGCGCAACATCGT
H37rv-glnA    AGTAACGTCTGCGCAACACGGGGTTGACTGACGGGCAATATCGG
bovis-glnA    AGTAACGTCTGCGCAACACGGGGTTGACTGACGGGCAATATCGG
avium-glnA    CGTAACGTGCGCGCAACATCGGGTTGACTGACGGGCAACATCTG

msmeg-amtB    GTTCACTTTCCGGAAACGCAACGGCAGCACCGGCGAAACGCCGG
H37rv-amtB    GTTAATCCTGATGAAACATGGCGGCACCATCGCCGCAACAACTA
bovis-amtB    GTTAATCCTGATGAAACATGGCGGCACCATCGCCGCAACAACTA
avium-amtB    GTTAATCCCGCCGAAACACAGCGGCACTATCGCCGAAACAACCA

msmeg-amt1    TTTAACCACGCTGCAACACTTGGCGACCATCTCCGTAACAGAAA
```

FIG. 2. GlnR binding motif in mycobacteria. Sequence logos of putative GlnR *cis* elements were identified upstream of the *M. smegmatis* (msmeg), *M. tuberculosis* (H37rv), *M. bovis* (bovis), and *M. avium* (avium) *glnA*, *amtB*, and *amt1* genes (*amt1* is present only in the *M. smegmatis* genome). The standard code of the Weblogo server is shown in gray scale at the top.

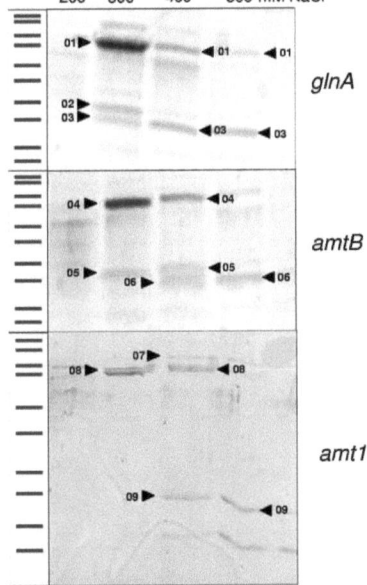

FIG. 3. DNA affinity purification of GlnR. Magnetic beads coated with *glnA*, *amtB*, and *amt1* promoter DNA were incubated with *M. smegmatis* cytoplasmic proteins. After washing steps, proteins bound to the DNA fragments were eluted using buffers containing different NaCl concentrations. Eluted proteins were separated by SDS-PAGE and subjected to tryptic in-gel digestion, MALDI-TOF MS, and peptide mass fingerprint analyses. The numbers indicate proteins identified by this approach, as shown in Table 1.

Middlebrook 7H9 medium, no difference between the wild-type and mutant strains was observed. The doubling times were between 4.5 and 5 h (data not shown). The response to nitrogen deprivation (i.e., the response after transfer of the bacteria to nitrogen-free Middlebrook 7H9 medium) was analyzed by performing RNA hybridization experiments. The transcription profiles of the *M. smegmatis* wild-type strain and ΔglnR strain MH1 were compared for the *glnA*, *amtB*, and *amt1* genes. As expected based on the results of the bioinformatic analyses and DNA affinity purification experiments described above, the transcript levels of these genes increased during nitrogen deprivation, while no increase in the level of transcription of *glnA*, *amtB*, and *amt1* was observed in *glnR* mutant strain MH1. This suggests that GlnR is in fact a nitrogen-dependent regulator in *M. smegmatis* and also indicates that GlnR works as an activator of transcription (Fig. 4A). The putative expression regulation of GlnR itself was tested by performing RNA hybridization using a *glnR* probe (Fig. 4B). Due to deletion of *glnR*, strain MH1 showed no *glnR* transcription signal, while the wild-type samples exhibited constitutive expression, which was not significantly altered upon nitrogen deprivation. This result indicates that *glnR* transcription is not subject to nitrogen control.

The regulation of mRNA levels in response to nitrogen starvation was also analyzed by performing real-time RT-PCR experiments. For the wild type (Fig. 5), these experiments showed that, compared to the levels after growth in Middlebrook 7H9 medium, the levels of transcripts were significantly increased after the washing step in Middlebrook 7H9 medium without a nitrogen source by factors of 50% ± 1% for *glnA*, 72% ± 3% for *amtB*, and 56% ± 3% for *amt1*. After 30 min the transcript levels were increased by 64% ± 4% for *glnA*, 340% ± 35% for *amtB*, and 537% ± 95% for *amt1*. For all time points, the upregulation factor was significantly higher for the ammonium transporter-encoding genes *amtB* and *amt1* than for *glnA*. Again, regulation of *glnR* was not observed (data not shown). For Δ*glnR* mutant MH1 no increases in mRNA levels in response to nitrogen starvation were observed. In fact, when the steady-state level for the wild type (zero time) was

TABLE 1. Proteins isolated by DNA affinity purification[a]

Accession no.	Protein	No.	Mol wt	Sequence coverage (%)	Promoter DNA
MSMEG_4954	Transcription termination factor Rho	01	71,762	60	glnA
MSMEG_3081	Hypothetical protein	02	34,499	71	glnA
MSMEG_5784	GlnR	03	27,933	39–60	glnA
MSMEG_4954	Transcription termination factor Rho	04	71,762	60	amtB
MSMEG_3081	Hypothetical protein	05	34,499	62–71	amtB
MSMEG_5784	GlnR	06	27,933	47–54	amtB
MSMEG_6157	DNA topoisomerase I	07	102,137	46	amt1
MSMEG_4954	Transcription termination factor Rho	08	71,762	37	amt1
MSMEG_5784	GlnR	09	27,933	26	amt1

[a] Identification of proteins after DNA affinity purification was carried out by performing tryptic in-gel digestion, MALDI-TOF MS, and peptide mass fingerprint analyses. The numbers are the same as the numbers in Fig. 3.

defined as 100%, the levels of transcription in the mutant were 19% ± 2% for glnA, 12% ± 1% for amtB, and 31% ± 1% for amt1, which is consistent with an activator function of GlnR. Moreover, glutamine synthetase activity was reduced in the mutant. While 0.230 ± 0.016 U (mg protein)$^{-1}$ was observed in the wild type, only 0.148 ± 0.009 U (mg protein)$^{-1}$ was observed in strain MH1.

Induction of GlnR-dependent transcription response by MSX treatment. As an alternative approach for inducing nitrogen starvation without changing the medium, the glutamate analog MSX was tested. MSX is known to inactivate glutamine synthetase and impair ammonium assimilation (15, 16). The growth tests carried out showed that addition of MSX to a final concentration of 200 μM did not impair growth of the wild type (doubling time, 4.5 to 5.0 h), while glnR mutant MH1 stopped growing after approximately 4 h of MSX treatment. At the mRNA level, within 15 min after addition of MSX enhanced transcription of glnA and amtB was observed in the wild type, while an increase in amt1 transcription was detected after 60 min (Fig. 6A), indicating that MSX in fact induces the cellular nitrogen starvation response at the level of transcription. As expected, this response at the level of transcription of glnA, amtB, and amt1 was not observed in glnR mutant strain MH1. Expression regulation of GlnR itself was also tested by

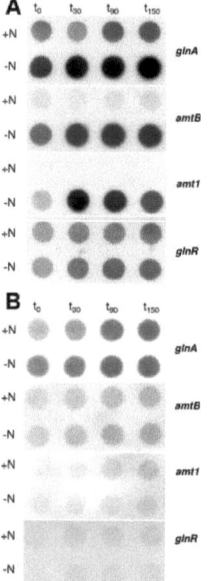

FIG. 4. Transcriptional responses of the M. smegmatis wild-type strain and glnR deletion strain MH1 to nitrogen deprivation: hybridization of RNA isolated from M. smegmatis cells before and after induction of nitrogen starvation by incubation in nitrogen-free medium. (A) Transcription of the glutamine synthetase-encoding glnA gene, amtB and amt1 coding for ammonium uptake systems, and glnR in the wild type. The time points indicated (t_0, t_{30}, t_{90}, and t_{150}) are sampling times (in min) after resuspension of centrifuged and washed cells. One microgram of total RNA per spot was applied. (B) Transcription of genes in M. smegmatis glnR deletion strain MH1.

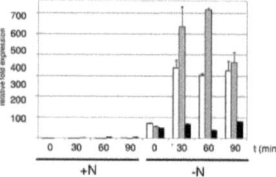

FIG. 5. Real-time RT-PCR of the M. smegmatis wild-type strain. Cells were grown until the exponential growth phase was reached (OD$_{600}$, approximately 0.6 to 0.8). To induce nitrogen starvation, cells were harvested, washed, and resuspended in prewarmed Middlebrook 7H9 medium without a nitrogen source (−N); prewarmed standard Middlebrook 7H9 medium was used as a control (+N). RNA was isolated at the indicated time points, and transcription of amtB (open bars), amt1 (gray bars), and glnA (black bars) was monitored by real-time RT-PCR. For each gene the control value at zero time was defined as 1.

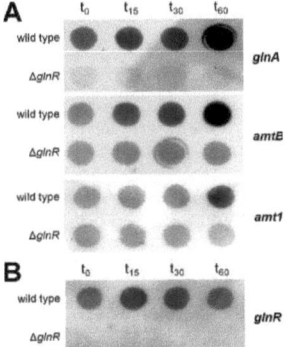

FIG. 6. Induction of the nitrogen starvation response by MSX: hybridization of RNA isolated from *M. smegmatis* cells before (t_0) and after addition of MSX. t_{15}, 15 min after addition of MSX; t_{30}, 30 min after addition of MSX; t_{60}, 60 min after addition of MSX. (A) Transcription of the glutamine synthetase-encoding *glnA* gene, as well as *amtB* and *amt1* coding for ammonium uptake systems. (B) Transcription of *M. smegmatis glnR*.

RNA hybridization using a *glnR* probe (Fig. 6B). As expected, wild-type samples exhibited constitutive expression of *glnR*, which was not significantly altered upon addition of MSX, confirming that *glnR* transcription is not subject to nitrogen control.

Complementation of strain MH1 by *M. smegmatis glnR*. To exclude the possibility of a polar effect of the *glnR* deletion, a complementation experiment was carried out. To do this, strain MH1 was transformed with plasmid pMN016 (control) and with the *glnR*-carrying plasmid pMN016-*glnR* (Fig. 7). Transcription of *glnR* was tested as a control (Fig. 7A). While the control plasmid without the *glnR* gene had no influence on transcription of *glnA*, *amtB*, and *amt1*, transcription of these genes in response to nitrogen starvation was increased as it was in the wild type when *glnR* was supplied in *trans* by plasmid pMN016-*glnR* (Fig. 7B). Consequently, the lack of nitrogen control in strain MH1 was caused exclusively by the *glnR* deletion and could be cured by complementation with a plasmid-encoded *glnR* gene.

Gel retardation assays. DNA affinity purification revealed that GlnR is able to bind to the promoter regions of *glnA*, *amtB*, and *amt1*. As an independent approach to verify GlnR binding, gel retardation experiments were carried out using corresponding promoter DNAs and cell extracts from the *M. smegmatis* wild-type strain and *glnR* deletion mutant MH1. As expected, a promoter DNA fragment of the *amtB* gene was shifted by the wild-type cell extract, while identical amounts of MH1 proteins had no influence on DNA mobility (data not shown). Nevertheless, these experiments with cell extracts did not exclude the possibility of indirect effects of GlnR on DNA motility. Therefore, GlnR was purified and used as an isolated protein in gel retardation experiments (Fig. 8). Again, the

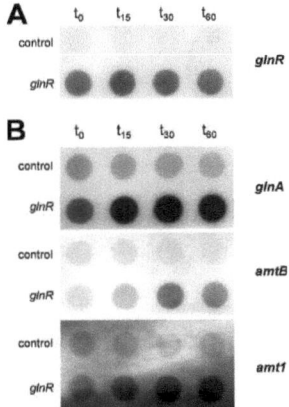

FIG. 7. Complementation of *M. smegmatis glnR* deletion strain MH1: hybridization of RNA isolated from *M. smegmatis* cells carrying control plasmid pMN016 or *glnR* delivery plasmid pMN016-*glnR* before (t_0) and after induction of nitrogen starvation by addition of MSX. t_{15}, 15 min after addition of MSX; t_{30}, 30 min after addition of MSX; t_{60}, 60 min after addition of MSX. (A) Transcription control for *M. smegmatis glnR*. (B) Transcription of *glnA*, *amtB*, and *amt1*.

promoter regions of *amtB* and *glnA* exhibited decreased motility in the gel when GlnR protein was added (Fig. 8A and B), showing that this transcriptional regulator binds to the corresponding DNA. To exclude the possibility of nonspecific, general DNA binding activity of GlnR, the corresponding *glnR* promoter DNA was tested. As shown by the RNA hybridization experiments described above (Fig. 4 and 6), *glnR* expression was not autoregulated, and in fact, no binding of GlnR was observed (Fig. 8C). Furthermore, promoter regions of urease-encoding genes (*ureE1* [msmeg_1091] and *ureA2* [msmeg_3627]) were tested, and no retardation was observed for these DNA fragments (data not shown). Together, these data indicate that there was specific binding of GlnR to the

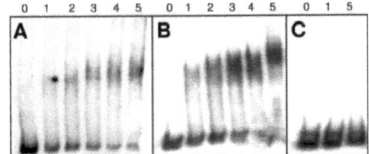

FIG. 8. Gel retardation assay. *amtB* (220 bp) (A), *glnA* (219 bp) (B), and *glnR* (249 bp) (C) apparent promoter fragments were incubated with different amounts of purified GlnR protein. Lanes 0, 1, 2, 3, 4, and 5 contained 0, 1, 2, 3, 4, and 5 µl of a 0.3-µg µl^{-1} solution, respectively.

promoter regions of genes, showing that there was GlnR-dependent upregulation of transcription upon nitrogen deprivation.

DISCUSSION

Previous work on nitrogen metabolism and its regulation in mycobacteria concentrated strongly on glutamine synthetase in *M. tuberculosis*, which is essential in this bacterium and consequently an important drug target (11, 12, 27). Additionally, studies of the *glnE* gene product adenylyltransferase, which is crucial for posttranslational modification and regulation of glutamine synthetase, and the *glnD*-encoded uridylyltransferase, which is putatively involved in nitrogen signal transduction, were carried out (11, 12, 18, 20). In this paper, we describe the first in vivo characterization of GlnR as a regulator of nitrogen control in mycobacteria. In *M. smegmatis*, GlnR binding sites were identified upstream of the *glnA*, *amtB*, and *amt1* genes. As in other actinomycetes (14), the *M. smegmatis amtB* gene forms an operon with downstream *glnK* and *glnD* genes. A core ammonium assimilation regulon seems to be controlled by *M. smegmatis* GlnR and to regulate transcription of *glnA*, which encodes glutamine synthetase, the ammonium transporter-encoding genes *amtB* and *amt1*, and corresponding signal transduction components encoded by *glnK* and *glnD*. The signal transduction to *M. smegmatis* GlnR is unclear, a situation which is similar to that in *S. coelicolor*. Based on the conserved phosphorylation domain and analogous to the situation in *E. coli* (32), involvement of a protein kinase is assumed. However, the corresponding protein is unknown.

Based on the data obtained for *M. smegmatis* in this study and the data obtained in previous studies focusing on *S. coelicolor* (8, 26, 29, 30) and *Amycolatopsis mediterranei* (33, 34), GlnR seems to be a major regulator of ammonium assimilation in actinomycetes. This conclusion is supported by the results of electrophoretic mobility shift assays, which showed that *S. coelicolor* GlnR is able to bind to the *glnA* promoter regions of *Bifidobacterium longum*, *Frankia* sp. strain EAN1, *M. tuberculosis*, *Nocardia farcinica*, *Nocardioides* sp. strain JS614, *Propionibacterium acnes*, and *Rhodococcus* sp. strain RHA1 in vitro (26). In contrast to the situation in most other actinomycetes, GlnR does not play a role in nitrogen control in the genus *Corynebacterium*. In this genus, no GlnR homologs were observed; instead, AmtR is the central nitrogen control protein (28).

Compared to the recently described extended *S. coelicolor* GlnR regulon (26), *M. smegmatis* GlnR has a reduced number of target genes. For the NADP-dependent glutamate dehydrogenase gene *gdhA* (msmeg_5442), the urease operon (msmeg_2627 to msmeg_3622), and the nitrite reductase genes (msmeg_0427 and msmeg_0428), genes that are under the control of GlnR in *S. coelicolor*, no GlnR *cis* elements were found in this study. In fact, preliminary data indicate that there may be nitrogen-dependent regulation of the nitrite reductase and urease (T. Bräu, personal communication), and AmtR might be an interesting candidate for a second nitrogen regulator, considering the presence of a gene encoding an AmtR homolog in the genome of *M. smegmatis* and the fact that for *C. glutamicum* the urease-encoding genes are under the control of the AmtR repressor (2). Together with *N. farcinica*, *Rhodococcus* sp. strain RHA1, and *Arthrobacter aurescens*, *M. smegmatis* seems to be one of the few actinomycetes besides the members of the corynebacterium family to have an AmtR homolog. Thus, future studies should focus on the role of AmtR in *M. smegmatis* using comparisons of transcriptome data for wild-type and deletion mutant strains.

ACKNOWLEDGMENTS

We thank D. Hillmann (Erlangen, Germany) for providing *M. smegmatis* strains, M. Niederweis (Birmingham, AL) for providing plasmids, and A. Lüdke (Erlangen, Germany) for help with protein analysis.

A.B., J.A., K.H., and F.T. were supported by the Deutsche Forschungsgemeinschaft (grant SFB 473), and J.S. received a fellowship from the Deutscher Akademischer Austausch Dienst.

REFERENCES

1. **Baumbach, J.** 2007. CoryneRegNet 4.0—a reference database for corynebacterial gene regulatory networks. BMC Bioinformatics **8:**429.
2. **Beckers, G., J. Strösser, U. Hildebrandt, J. Kalinowski, M. Farwick, R. Krämer, and A. Burkovski.** 2005. Regulation of AmtR-controlled gene expression in *Corynebacterium glutamicum*: mechanism and characterization of the AmtR regulon. Mol. Microbiol. **58:**580–595.
3. **Belisle, J. T., and M. G. Sonnenberg.** 1998. Isolation of genomic DNA from mycobacteria. Methods Mol. Biol. **101:**31–44.
4. **Burkovski, A.** 2003. Ammonium assimilation and nitrogen control in *Corynebacterium glutamicum* and its relatives: an example for new regulatory mechanisms in actinomycetes. FEMS Microbiol. Rev. **27:**617–628.
5. **Burkovski, A.** 2005. Nitrogen metabolism and its regulation, p. 333–349. *In* M. Bott and L. Eggeling (ed.), Handbook of *Corynebacterium glutamicum*. CRC Press LLC, Boca Raton, FL.
6. **Burkovski, A.** 2007. Nitrogen control in *Corynebacterium glutamicum*: proteins, mechanisms, signals. J. Microbiol. Biotechnol. **17:**187–194.
7. **Carroll, P., C. A. Pashley, and T. Parish.** 2008. Functional analysis of GlnE, an essential adenylyl transferase in *Mycobacterium tuberculosis*. J. Bacteriol. **190:**4894–4902.
8. **Fink, D., N. Weisschuh, J. Reuther, W. Wohlleben, and A. Engels.** 2002. Two transcriptional regulators GlnR and GlnRII are involved in regulation of nitrogen metabolism in *Streptomyces coelicolor* A3(2). Mol. Microbiol. **46:**331–347.
9. **Hänßler, E., and A. Burkovski.** 2008. Molecular mechanisms of nitrogen control in corynebacteria, p. 183–201. *In* A. Burkovski (ed.), Corynebacteria: genomics and molecular biology. Caister Academic Press, Wymondham, United Kingdom.
10. **Harper, C., D. Hayward, I. Wild, and P. Van Helden.** Regulation of nitrogen metabolism in *Mycobacterium tuberculosis*: a comparison with mechanisms in *Corynebacterium glutamicum* and *Streptomyces coelicolor*. IUBMB Life, in press.
11. **Harth, G., D. L. Clemens, and A. A. Horwitz.** 1994. Glutamine synthetase of *Mycobacterium tuberculosis*: extracellular realease and characterization of its enzymatic activity. Proc. Natl. Acad. Sci. USA **91:**9342–9346.
12. **Harth, G., S. Maslesa-Galic, M. V. Tullius, and A. A. Horwitz.** 2005. All four *Mycobacterium tuberculosis glnA* genes encode glutamine synthetase activities but only GlnA1 is abundantly expressed and essential for bacterial homeostasis. Mol. Microbiol. **58:**1157–1172.
13. **Hiard, S., R. Maree, S. Colson, P. A. Hoskisson, F. Titgemeyer, G. van Wezel, B. Joris, L. Wehenkel, and S. Rigali.** 2007. PREDetector: a new tool to identify regulatory elements in bacterial genomes. Biochem. Biophys. Res. Commun. **357:**861–864.
14. **Jakoby, M., L. Nolden, J. Meier-Wagner, R. Krämer, and A. Burkovski.** 2000. AmtR, a global repressor in the nitrogen regulation system of *Corynebacterium glutamicum*. Mol. Microbiol. **37:**964–977.
15. **Khan, A., S. Akhtar, J. Ahmad, and D. Sarkar.** 2008. Presence of a functional nitrate pathway in *Mycobacterium smegmatis*. Microb. Pathog. **44:**71–77.
16. **Manning, J. M., S. Moore, W. B. Rowe, and A. Meister.** 1969. Identification of L-methionine S-sulfoximine as the diastereoisomer of L-methionine SR-sulfoximine that inhibits glutamine synthetase. Biochemistry **8:**2681–2686.
17. **Nolden, L., G. Beckers, and A. Burkovski.** 2002. Nitrogen assimilation in *Corynebacterium diphtheriae*: pathways and regulatory cascades. FEMS Microbiol. Lett. **208:**287–293.
18. **Parish, T., and N. G. Stoker.** 2000. *glnE* is an essential gene in *Mycobacterium tuberculosis*. J. Bacteriol. **182:**5715–5720.
19. **Pashley, C. A., A. C. Brown, D. Robertson, and T. Parish.** 2006. Identification of the *Mycobacterium tuberculosis glnH* promoter and its response to nitrogen availability. Microbiology **152:**2727–2734.
20. **Read, R., C. A. Pashley, D. Smith, and T. Parish.** 2007. The role of GlnD in ammonium assimilation in *Mycobacterium tuberculosis*. Tuberculosis **87:**384–390.

21. **Reuther, J., and W. Wohlleben.** 2007. Nitrogen metabolism in *Streptomyces coelicolor*: transcriptional and post-translational regulation. J. Mol. Microbiol. Biotechnol. **12:**139–146.
22. **Sambrook, J., E. F. Fritsch, and T. Maniatis.** 1989. Molecular cloning: a laboratory manual, 2nd ed. Cold Spring Harbor Laboratory, Cold Spring Harbor, NY.
23. **Schägger, H., and G. von Jagow.** 1987. Tricine-sodium dodecyl sulfate-polyacrylamide gel electrophoresis for the separation of proteins in the range from 1 to kDa. Anal. Biochem. **166:**368–379.
24. **Stephan, J., V. Stemmer, and M. Niederweis.** 2004. Consecutive gene deletions in *Mycobacterium smegmatis* using the yeast FLP recombinase. Gene **343:**181–190.
25. **Stephan, J., J. Bender, F. Wolschendorf, C. Hoffmann, E. Roth, C. Mailänder, H. Engelhardt, and M. Niederweis.** 2005. The growth rate of *Mycobacterium smegmatis* depends on sufficient porin-mediated influx of nutrients. Mol. Microbiol. **58:**714–730.
26. **Tiffert, Y., P. Supra, R. Wurm, W. Wohlleben, R. Wagner, and J. Reuther.** 2008. The *Streptomyces coelicolor* GlnR regulon: identification of new GlnR targets and evidence for a central role of GlnR in nitrogen metabolism in actinomycetes. Mol. Microbiol. **67:**436–446.
27. **Tullius, M. V., G. Harth, and A. A. Horwitz.** 2003. Glutamine synthetase GlnA1 is essential for growth of *Mycobacterium tuberculosis* in human THP-1 macrophages and guinea pigs. Infect. Immun. **71:**3927–3936.
28. **Walter, B., E. Hänßler, J. Kalinowski, and A. Burkovski.** 2007. Nitrogen metabolism and nitrogen control in corynebacteria: variations of a common theme. J. Mol. Microbiol. Biotechnol. **12:**131–138.
29. **Wray, L. V., and S. H. Fisher.** 1993. The *Streptomyces coelicolor glnR* gene encodes a protein similar to other bacterial response regulators. Gene **130:**145–150.
30. **Wray, L. V., M. R. Atkinson, and S. H. Fisher.** 1991. Identification and cloning of the *glnR* locus, which is required for transcription of the *glnA* gene in *Streptomyces coelicolor* A3(2). J. Bacteriol. **173:**7351–7360.
31. **Yanisch-Perron, C., L. Vieira, and J. Messing.** 1985. Improved M13 phage cloning vectors and host strains: nucleotide sequences of M13mp18 and pUC19 vectors. Gene **33:**103–119.
32. **Yoshida, T., L. Qin, L. A. Egger, and M. Inouye.** 2006. Transcription regulation of *ompF* and *ompC* by a single transcription factor, OmpR. J. Biol. Chem. **281:**17114–17123.
33. **Yu, H., W. Peng, Y. Liu, T. Wu, Y. Yao, M. Cui, W. Jiang, and G.-P. Zhao.** 2006. Identification and characterization of *glnA* promoter and its corresponding *trans*-regulating protein GlnR in the rifamycin SV producing actinomycete, Amycolatopsis mediterranei U32. Acta Biochim. Biophys. Sin. **38:**831–843.
34. **Yu, H., Y. Yao, Y. Liu, R. Jiao, W. Jiang, and G.-P. Zhao.** 2007. A complex role of *Amycolatopsis mediterranei* GlnR in nitrogen metabolism and related antibiotics production. Arch. Microbiol. **188:**89–96.

A Genomic View of Sugar Transport in *Mycobacterium smegmatis* and *Mycobacterium tuberculosis*[▽]

Fritz Titgemeyer,[1]* Johannes Amon,[1] Stephan Parche,[1] Maysa Mahfoud,[1] Johannes Bail,[1] Maximilian Schlicht,[1] Nadine Rehm,[1] Dietmar Hillmann,[1,2] Joachim Stephan,[1] Britta Walter,[1] Andreas Burkovski,[1] and Michael Niederweis[1,2]*

Lehrstuhl für Mikrobiologie, Friedrich Alexander Universität Erlangen-Nürnberg, Staudtstr. 5, D-91058 Erlangen, Germany,[1] and Department of Microbiology, University of Alabama at Birmingham, 613 Bevill Biomedical Research Building, 845 19th Street South, Birmingham, Alabama 35294[2]

Received 14 February 2007/Accepted 22 May 2007

> We present a comprehensive analysis of carbohydrate uptake systems of the soil bacterium *Mycobacterium smegmatis* and the human pathogen *Mycobacterium tuberculosis*. Our results show that *M. smegmatis* has 28 putative carbohydrate transporters. The majority of sugar transport systems (19/28) in *M. smegmatis* belong to the ATP-binding cassette (ABC) transporter family. In contrast to previous reports, we identified genes encoding all components of the phosphotransferase system (PTS), including permeases for fructose, glucose, and dihydroxyacetone, in *M. smegmatis*. It is anticipated that the PTS of *M. smegmatis* plays an important role in the global control of carbon metabolism similar to those of other bacteria. *M. smegmatis* further possesses one putative glycerol facilitator of the major intrinsic protein family, four sugar permeases of the major facilitator superfamily, one of which was assigned as a glucose transporter, and one galactose permease of the sodium solute superfamily. Our predictions were validated by gene expression, growth, and sugar transport analyses. Strikingly, we detected only five sugar permeases in the slow-growing species *M. tuberculosis*, two of which occur in *M. smegmatis*. Genes for a PTS are missing in *M. tuberculosis*. Our analysis thus brings the diversity of carbohydrate uptake systems of fast- and a slow-growing mycobacteria to light, which reflects the lifestyles of *M. smegmatis* and *M. tuberculosis* in their natural habitats, the soil and the human body, respectively.

The growth and nutritional requirements of mycobacteria have been intensely studied since the discovery of *Mycobacterium tuberculosis* (32). This resulted in an overwhelming body of literature on the physiology of mycobacterial metabolism in the years before the dawn of molecular biology (20, 53, 54). Carbon metabolism of mycobacteria has attracted renewed interest since the discovery that *M. tuberculosis* relies on the glyoxylate cycle for survival in mice (36, 41). This observation indicates that *M. tuberculosis* uses lipids as the main carbon source during infection. On the other side, genes that encode a putative disaccharide transporter were essential for *M. tuberculosis* during the first week of infection, indicating that *M. tuberculosis* may switch its main carbon source from carbohydrates to lipids with the onset of the adaptive immune response (61). However, the nutrients and the corresponding uptake proteins are unknown for *M. tuberculosis* inside the human host. Surprisingly, this is also true for *M. tuberculosis* growing in vitro and for *Mycobacterium smegmatis*, which is often used as a fast-growing, nonpathogenic model organism to learn more about basic mycobacterial physiology. There is no doubt that the uptake pathways have been adapted to the habitats of *M. tuberculosis* and *M. smegmatis*, the human body and soil, respectively. Thus, much can be learned about the lifestyles of both organisms by a comparison of the complements of specific nutrient uptake proteins. Previously, 38 ATP-binding cassette (ABC) transport proteins have been identified in *M. tuberculosis* by bioinformatic analysis, 4 of which were assigned a role in carbohydrate import (9).

The goal of this study was to compile a complete list of potential carbohydrate uptake systems of *M. smegmatis* and *M. tuberculosis* based on in silico analyses of their genomes. While *M. tuberculosis* has only 5 recognizable carbohydrate import systems, *M. smegmatis* has 28 of such transporters at its disposal. In particular, we show that the genome of *M. smegmatis* encodes a phosphotransferase system (PTS) that plays a fundamental role in the global control of sugar metabolism in both gram-negative and gram-positive bacteria (12). Furthermore, we show by reverse transcription (RT)-PCR and uptake experiments that the PTS genes and other selected systems are functionally expressed in *M. smegmatis*.

MATERIALS AND METHODS

Chemicals and enzymes. Chemicals were purchased from Merck, Roth, or Sigma at the highest purity available. Enzymes for DNA restriction and modification were obtained from New England Biolabs, MBI Fermentas, and Boehringer. Oligonucleotides were obtained from MWG-Biotech AG.

Bacterial strains and growth conditions. *M. smegmatis* mc²155 was grown in liquid cultures using Middlebrook 7H9 medium (Difco) supplemented with 0.2% glycerol and 0.05% Tween 80 or minimal Hartmans-de Bont (HB) medium (65)

---

* Corresponding author. Mailing address for Fritz Titgemeyer: Lehrstuhl für Mikrobiologie, Friedrich Alexander Universität Erlangen-Nürnberg, Staudtstr. 5, D-91058 Erlangen, Germany. Phone: 49 177 4824821. Fax: 49 9131 8528082. E-mail: fritz.titgemeyer@googlemail.com. Mailing address for Michael Niederweis: Department of Microbiology, University of Alabama at Birmingham, 613 Bevill Biomedical Research Building, 845 19th Street South, Birmingham, AL 35294. Phone: (205) 996 2711. Fax: (205) 934 9256. E-mail: mnieder@uab.edu.

▽ Published ahead of print on 8 June 2007.

at 37°C. *M. smegmatis* mc²155 was grown on plates using Middlebrook 7H10 medium supplemented with 0.5% glycerol.

Computer analyses and screening strategies. Protein sequences of known carbohydrate uptake systems were used to screen the genome sequence of *M. smegmatis* mc²155 at the BLAST server of the National Center for Biotechnology Information (NCBI) at the National Institutes of Health, Bethesda, MD (www.ncbi.nlm.nih.gov/sutils/genom_table.cgi), using TBLASTN. A data file containing the preliminary genome sequence of *M. smegmatis* mc²155 containing 7,278,076 nucleotides was obtained from The Institute for Genomic Research (TIGR) (www.tigr.org). This file was loaded into the ARTEMIS software available at the Wellcome Trust Sanger Institute (www.sanger.ac.uk) to annotate gene and protein sequences. The open reading frames (ORFs) and their adjacent genes were checked by visual inspection to detect the most likely start codon that is preceded by a ribosome binding site. Sizes of ORFs were further established by a consideration of codon bias analysis as implemented in Artemis and by multiple sequence alignments with well-characterized homologs. For the latter, the *M. smegmatis* ORFs were subjected to general PBLAST data bank searches at the NCBI website (www.ncbi.nlm.nih.gov) to detect closely related sequences. Finally, the identified genes were cross-checked with the primary annotation protein list (http://cmr.tigr.org). To find the most representative homologs, we used single genome protein BLAST, which is available for well-characterized bacterial species of diverse phylogenetic origins: *Escherichia coli*, *Bacillus subtilis*, *M. tuberculosis* (all accessible at http://genolist.pasteur.fr), and *Streptomyces coelicolor* (http://avermitilis.ls.kitasato-u.ac.jp). Prediction of the possible substrate(s) was based on the following criteria: (i) protein identity of the *M. smegmatis* protein was more than 30% to the homologous protein of *E. coli* or *B. subtilis*, more than 50% to *Streptomyces*, or more than 60% to *M. tuberculosis*; (ii) more than one gene of the operon was conserved; and, most importantly, (iii) a solid biochemical analysis of the homologous protein was available. Sequence alignments were conducted with CLUSTALW by applying predefined algorithms available from the European Bioinformatics Institute at The European Molecular Biology Laboratory (www.ebi.ac.uk/clustalw).

Growth of *M. smegmatis* on sugars. A 4-ml culture of *M. smegmatis* mc²155 was grown in HB minimal medium supplemented with 0.05% Tween 80 and 1% sugars as the single carbon source (65). Cells were passed through a filter with a pore size of 5 μm to remove cell clumps and were then inoculated into 50 ml HB medium. The 50-ml cultures were grown until an optical density at 600 nm (OD₆₀₀) of 1 was reached. The bacteria were harvested at 4,000 rpm at 4°C for 10 min and washed twice with minimal medium without a carbon source. The pellet was resuspended in 5 ml HB medium and diluted to an OD₆₀₀ of 0.02 in 50 ml HB medium containing 1% carbon source. Growth rates were determined in three independent cultures by OD₆₀₀ measurements every 3 h.

Reverse transcriptase PCR. Cells of *M. smegmatis* mc²155 were grown in HB minimal medium with 50 mM glycerol. In other cultures, 50 mM of the carbon source of interest was added to examine gene-specific induction. The cultures were harvested at mid-exponential phase and subjected to total RNA preparation using a procedure that we described previously (68). Quantitative RT-PCR experiments were conducted as described previously (73). RT-PCRs were performed with gene-specific oligonucleotides, which were AACTGTGCTTTCTC AACCG and ATGGCGTCGAGTTGGTGC for *ptsI*, TCACCGTCGGATCTG CCGTCG and ACCAGTTCGGCAACCTTGGC for *ptsH*, AGGCATCAACG TGGCAAGG and ACCGCGTGATCGCATCGAGCG for *fruK*, ACCGAGT TCCTGTTCCTCG and CGAGCGTCGTGACCATCG for *ptsI*, GGGCATCC TCACGTCAGG and CAGCAGGTCGATCAGACC for *ptsG*, TCAGACCGT GACCATCACG and TGGAACCAGCACTCCCAC for *crr*, and GCAAGGTGC TTCCGTTCAGC and CGAGACCGATGATCACCG for *glpF1*. The assay mixture contained 100 ng of RNA and 5 pmol of each primer in a 20-μl volume. Samples of 4 μl of each reaction mixture were taken at appropriate PCR cycles (cycles 21 to 36 depending on the appearance of a signal in the linear range), and amplification products were separated and visualized on a 1% agarose gel. RT-PCR experiments without prior RT were performed to ensure the exclusion of DNA contamination. The quality of the RNA preparations was checked by the presence of equal amounts of 16S rRNA, which is constitutively expressed. Data were verified in two independent experiments.

Transport assays. Sugar uptake measurements were carried out as previously described (67). To reduce aggregation and clumping, all *M. smegmatis* cells were filtered through a 5-μm-pore-size filter (Sartorius) and regrown for 2 days at 37°C before inoculating 100-ml cultures (69). Cells were grown in the presence of 0.2 or 0.4% of the respective carbon source as the standard carbon source and harvested by centrifugation (1,250 × *g* at 4°C for 10 min) when they had reached the mid-exponential phase at an OD₆₀₀ of between 0.5 and 0.7. The cells were washed once in 2 mM PIPES [piperazine-*N*,*N*'-bis(2-ethanesulfonic acid)] (pH 6.5)–0.05 mM MgCl₂ and resuspended in the same buffer. Radiolabeled [¹⁴C]fructose, [¹⁴C]glucose, and [¹⁴C]glycerol were added to the cell suspension to obtain final sugar concentrations of 20 μM, 20 μM, and 100 μM, respectively, and a radioactivity of 100,000 cpm per ml cells. The mixtures were incubated at 37°C for glycerol uptake and 25°C for glucose and fructose uptake assays. Cell suspensions (between 0.2 and 1 ml) were filtered through a 0.45-μm-pore-size filter (Sartorius) and washed with 0.1 M LiCl, and the radioactivity was determined by using a liquid scintillation counter (Beckman). All experiments were reproduced by at least one biological replicate.

RESULTS

Genome analysis reveals 28 putative carbohydrate uptake systems in *M. smegmatis*. The sequence of the almost finished genome of *M. smegmatis* mc²155 was searched for homologs of well-characterized bacterial sugar transport systems to identify possible carbohydrate uptake systems. Table 1 shows a list of 28 putative carbohydrate permeases: 19 of the ABC family, 3 of the PTS family, 1 of the major intrinsic protein family (MIP), 4 of the major facilitator superfamily (MFS), and one of the sodium solute superfamily (SSS) (6, 47–49, 56).

ABC systems. Operons for carbohydrate-specific transporters of the ABC family always contain a gene encoding a sugar-specific periplasmic binding protein (6, 9). We have conducted TBLASTN analyses of the *M. smegmatis* genome with known binding proteins of ABC systems, such as the maltose- and ribose-specific binding proteins MalE and RbsB, respectively, of *E. coli* and the cellobiose binding protein of *Streptomyces reticuli* (19, 20, 62). We found 18 genes for ABC-type sugar binding proteins in the genome of *M. smegmatis* mc²155. All were adjacent to ABC permease genes (Fig. 1 and Table 1). For all sugar transport systems, a substrate was predicted when the following criteria were met (6): (i) a solid biochemical analysis of the homologous proteins was available, (ii) the identity of the *M. smegmatis* protein to the homologous proteins of *E. coli* or *B. subtilis* was greater than 35%, and that to *Streptomyces* was greater than 35% and 50%, respectively, (iii) more than one gene of the operon was conserved, and (iv) the encoded proteins had the same carbohydrate specificity. In this way, we predict that *M. smegmatis* can transport via ABC permeases, β-glucosides such as chitobiose, α-galactosides (melibiose), β-xylosides (xylobiose), xylose, arabinose, and sugar alcohols. In addition, we propose that *M. smegmatis* has several ABC systems for ribose or ribose-like substrates such as ribonucleosides, ribitol, or xylitol (Table 1).

(i) **β-Glucosides (msmeg_0501 to msmeg_0508).** The first sugar transport cluster in the genome contains genes for a glucosamine isomerase (*nagB1*), a β-glucosidase (*bglA*), a sugar kinase (*sugK*), and a complete ABC permease, the latter with distant similarities to other ABC importers. Due to the presence of the metabolic genes, whose products BglA and NagB exhibit 31% and 32% protein identity to homologs of *B. subtilis* and *E. coli*, we suggest that the permease catalyzes the uptake of β-glucosides such as cellobiose (38, 72). It should be noted as well that the closest homologs were the *S. coelicolor* proteins SCO5236 (NagB homolog with 50% identity) and SCO6670 (BglA homolog with 54% identity).

(ii) **α-Galactosides (msmeg_0509 to msmeg_0517).** A gene cluster directly downstream of the putative β-glucoside cluster may be responsible for the uptake of α-galactosides. It comprises genes encoding an α-galactosidase (msmeg_0514), a tagatose-bisphosphate aldolase (*agaZ*), and a putative isomerase

TABLE 1. Sugar transport systems of *M. smegmatis*

Family and predicted substrate(s)	Gene designation(s)	Locus tag(s)[a]	Representative homolog(s)[b] or description	Reference(s) and/or source(s)[c]
ABC				
β-Glucosides, chitobiose, disaccharides	nagB1 bglA bglR sugK bglEFGK	0501–0508	nagB (SCO5236), bgl (SCO0670), deoR, rbsK E. coli; abcEFG genes are distantly related to many ABC permease genes; msiK (SCO4240)	6, S, C
α-Galactosides, melibiose	agaRZSXPAEFGK mspB	0509–0517	SCO5848–SCO5851, SCO0538–SCO0541, msiK (SCO4240)	6, 11, S
Unknown	abcEFGK	0553–0556	Distant similarity to many ABC permease genes	S, C
Ribose, xylose	rbsA1C1B1 gatY sugK rbsR1 pfkB	1372–1378	SCO6009–SCO6011, gatY (SCO5852), pfkB (SCO3197); distantly related to ribose and xylose ABC transporters	3, 6, S, C, T, B
Xylose	xylF2G2H2	1704–1706	E. coli xylFGH	17, 70, C
Arabinose	pho araR araGFKEBDA	1707–1715	E. coli ytf operon, B. subtilis araABD	60, B, C
Ribose, ribonucleosides	gap pgk tpiA secG urf rscA deoC rbsC2A2 rbsR2 sugK sugD rbsB2 ppc pgl opcA zwf tal tkt	3084–3103	rbsH (SCO2747), rbsA (SCO2746), E. coli rbsB	3, 6, S, C
Unknown	abcKGFE fabG sugK	3108–3113	Distant similarity to many ABC permease genes	6, S, C
Unknown	abcR sugK abcEFGK1K2 sugK rpiB	3264–3272	Distant similarity to many ABC permease genes, SCO0580, rpiB E. coli	6, 66, S, C
Ribose	rbsB3R3C3A3	3598–3602	SCO2747, B. subtilis ribose operon	75
Sorbitol	yphREKFB	3998–4002	yph operon E. coli	6, C
Ribose	rbsA4C4B4R4	4170–4174	E. coli ribose operon rbsCBR	3, C
Unknown	uspGFE	4466–4468	M. tuberculosis usp operon	9, T
Unknown	abcRFGE	4655–4658	Distant similar to many ABC permease genes	S, C
Unknown	sugKGFE	5058–5061	sug operon M. tuberculosis	9, T
β-Xyloside	bglG bxlRAEFG	5142–5147	S. coelicolor bxlEFG2	6
Sugar alcohol	smoKGFER	5571–5574	S. coelicolor smo operon	6
Xylose	xylG1F1E1A1R1	6018–6022	E. coli xyl operon	70
Ribose	rbsR5P5K5G5F5E5	6798–6805	SCO0723, SAV5702, SCO2747	6, S, C
PTS				
Fructose	ptsH fruAKR ptsI	0084–0088	S. coelicolor ptsH fruAKR ptsI	6, 45, S
Glucose, trehalose N-acetylglucosamine	ptsG crr nagB2A ptsR	2116–2120	S. coelicolor nagE2 crr nagAB	5, 44
Dihydroxyacetone	ptsT dhaLKFR	2121–2125	S. coelicolor gyl operon, E. coli dhaKLM	6, 25, S, C
MIP				
Glycerol	glpK2RFK1D	6756–6760	S. coelicolor gyl operon	6, S
SSS				
Galactose	galPRTK	3689–3692	S. coelicolor gal operon	6
MFS				
Unknown	sugP1	2966	Distant similarity to sugar permeases of the MFS	S, C, B, T
Unknown	sugP2	4078		
Glucose	glcP	4182	S. coelicolor glcP	73
Unknown	sugP3	5559		

[a] Only the numbers of the locus tags (msmeg_XXXX) of the *M. smegmatis* mc²155 genome are shown. Numbers are according to the revised annotation (www.tigr.org).
[b] The column contains information on representatives homologs for which experimental information is available.
[c] The following genome servers were used for BLASTP analysis: http://genolist.pasteur.fr/SubtiList (B) for *B. subtilis*, http://genolist.pasteur.fr/Colibri (C) for *E. coli*, www.avermitilis.ls.kitasato-u.ac.jp/blast_local/index.html (S) for *S. coelicolor*, and http://genolist.pasteur.fr/TubercuList (T) for *M. tuberculosis*.

(*agaS*) besides the genes for the ABC permease, *agaEFG*. The operon is likely controlled by AgaR, a regulator of the DeoR family (11, 51). Downstream of the *aga* region is the gene *mspB*, which encodes a porin (69), which may indicate that this porin is required for the entry of the substrate transported by AgaEFG. The proteins share, as many other proteins from *M. smegmatis* do, the highest identity to proteins from *S. coelicolor* (39 to 71%) (Table 1).

FIG. 1. Genetic organization of *M. smegmatis* carbohydrate transporters of the ABC family. The arrows indicate the lengths and transcriptional orientations of annotated genes and predicted ORFs. Genes encoding transport systems are depicted in dark gray, carbohydrate metabolic genes are colored in light gray, and regulatory genes are highlighted in black, while other genes are white. Genes are shown by their number, with the prefix "msmeg_." The gene names are assigned according to the annotations given by TIGR (http://www.tigr.org) and by us. Numbers in brackets refer to the intergenic distance between two genes. General gene designations are as follows: *urf*, unknown reading frame; *sugD*, sugar dehydrogenase; *sugK*, sugar kinase; *sugP*, sugar permease; *abcE*, unspecified substrate binding protein of an ABC permease; *abcF* and *abcG*, unspecified membrane proteins of an ABC permease.

(iii) **Ribose and ribose-like carbohydrates (msmeg_1372 to msmeg_1378, msmeg_3090 to msmeg_3095, msmeg_3598 to msmeg_3602, msmeg_4170 to msmeg_4174, and msmeg_6798 to msmeg_6804).** Several of the analyzed ABC systems of *M. smegmatis* revealed similarities to the ribose ABC permeases of *E. coli* and *B. subtilis* (3, 75). The best candidates for a ribose-specific ABC permease seem to be msmeg_3090, msmeg_3091, and msmeg_3095. The latter one is the substrate binding protein that shares the highest identity (30%) to the ribose-specific periplasmic binding protein RbsB. Interestingly, this putative ribose permease is embedded within a cluster of genes involved in central carbon metabolism, such as glycolysis. The other four ABC systems with similarities to ribose permeases may also transport ribose or a ribose derivative (Table 1).

(iv) **Xylose and β-xylosides (msmeg_1704 to msmeg_1706, msmeg_5142 to msmeg_5147, and msmeg_6018 to msmeg_6022).** Xylose is usually taken up by an ABC permease and further metabolized by isomerization (xylose isomerase [XylA]) and phosphorylation (xylulokinase [XylB]) to enter the pentose-phosphate shunt. The region msmeg_6018 to msmeg_6022 was designated the *xylGFEAR1* regulon since it encodes proteins sharing identities of 29% to 38% with the corresponding proteins of *E. coli* (7, 20). The substrate binding protein XylF1 has the highest score, with 38% identity to the *E. coli* homolog (70). The regulon further contains the metabolic *xylA1* gene. The missing XylB may be encoded by msmeg_3257, having 50% and 32% identity to XylB of *Corynebacterium glutamicum* and *E. coli*, respectively (31, 33). A second potential operon for a xylose ABC transporter is encoded by msmeg_1704 to msmeg_1706 (*xylFGH2*). XylF2, XylG2, and XylH2 share 38% to 48% identity to the corresponding *E. coli* proteins.

A predicted β-xyloside permease is encoded by msmeg_5142 to msmeg_5147. We have designated the genes *bxlRAEFG* according to their close relationship to the *bxlEFG2* operon of *S. coelicolor*, which encodes the transporter for xylobiose (6, 27). The amino acid identities are in the range of 45% to 79%.

(v) **Arabinose (msmeg_1708 to msmeg_1715).** We detected an *araBDA* operon that is most similar to the arabinose metabolic genes of *B. subtilis* (40). AraB (L-ribulokinase), AraD (L-ribulose-5-phosphate-4-epimerase), and AraA (L-arabinose isomerase) show 42 to 48% identity to the corresponding proteins of *B. subtilis*. The adjacent ABC permease (msmeg_1709 to msmeg_1712) could be the uptake system for arabinose. Its gene products, including the juxtaposed regulator, exhibit residues that are up to 43% identical to an unknown gene cluster of *E. coli* that is designated *ytfRT yjfF*. This suggests that the *E. coli* locus encodes a sugar-specific permease. Transport of arabinose has so far been described only for the *E. coli* proton symporter AraE in the MFS (34, 47).

(vi) **Sugar alcohols (msmeg_3998 to msmeg_4002 and msmeg_5571 to msmeg_5575).** Sugar alcohols like mannitol, glucitol (sorbitol), and xylitol are frequently consumed by bacteria (52). We found two loci that could encode a sugar alcohol-specific transport system. The operon comprising msmeg_3998 to msmeg_4002 is homologous to the *yph* operon of *E. coli*. Although the permease proteins do not allow substrate prediction, they are associated with a sugar alcohol dehydrogenase (msmeg_4002), which is 44% identical to YphC and more distantly related to GutB of *B. subtilis* (sorbitol dehydrogenase) and GatD (galactitol-1P dehydrogenase) of *E. coli* (43). The second region, msmeg_5571 to msmeg_5575, is homologous to the *smo* operon of *S. coelicolor*, encoding a possible permease for sugar alcohols. Here, the two substrate-specific binding SmoE proteins share 52% identity (6).

(vii) **Less-well-defined ABC-type sugar transporters.** We further found six gene loci encoding ABC transport systems for carbohydrates. The deduced substrate binding proteins were homologous to known sugar substrate binding proteins in the range of about 30% protein identity. Due to this and due to the absence of adjacent metabolic genes, prediction of substrates was not possible. Among those are the only two permeases that are common to *M. smegmatis* and *M. tuberculosis*, Sug^{ABC} and Usp^{ABC} (see below).

(viii) **Lipid anchors of sugar binding proteins.** In gram-negative bacteria, the substrate binding proteins of ABC transporters are soluble proteins that appear to move freely through the periplasmic space. By contrast, it has been shown for gram-positive bacteria that the substrate binding proteins are covalently anchored to the outside of the cell membrane by fatty acids (63). Lipidation occurs via esterification of a conserved cysteine at the N terminus of the processed protein. Strikingly, all sugar binding proteins of *M. smegmatis* (Table 2) and *M. tuberculosis* (9) have a predicted lipoprotein signal peptide.

PTS permeases. (i) Glucose, trehalose, GlcNAc, and dihydroxyacetone (msmeg_2116 to msmeg_2125). Two loci encoding components of a PTS were discovered in the genome of *M. smegmatis*. This is interesting, since the existence of PTS genes is in contrast to previous reports in which no evidence for PTS proteins in *M. smegmatis* was found in biochemical experiments (14, 15, 54, 59). Genes for a PTS of the glucose-sucrose subfamily (Fig. 2) were detected in a cluster comprising msmeg_2116 to msmeg_2120 (52). The two divergently transcribed genes msmeg_2116 (*ptsG*) and msmeg_2117 (*crr*) encode the IIBC and IIA permeases. The deduced protein sequence of IIBC (496 amino acids [aa]) shares 49% identity with the *N*-acetylglucosamine (GlcNAc)-specific IIB^{GlcNAc} and IIC^{GlcNAc} of *S. coelicolor* A3(2) and 42 to 43% identity to PTS permeases of *E. coli* and *B. subtilis* that transport glucose, GlcNAc, trehalose, or β-glucosides, respectively (52). The IIA protein of *M. smegmatis* is 50% identical to the *S. coelicolor* homolog IIA^{Crr} (30). Two metabolic genes for the metabolism of GlcNAc are situated downstream of the *crr* gene. They encode a putative glucosamine-6-phosphate deaminase, NagB2, and a GlcNAc deacetylase, NagA, that share 43% and 33% identity to the characterized *E. coli* homologs and 59% and 47% to the ones from *S. coelicolor*, respectively (1). The gene locus further contains a putative regulatory gene, *ptsR*, whose product shares low similarities to uncharacterized regulators of *S. coelicolor* and *M. tuberculosis*. Thus, PtsR might be a possible regulator of the PTS genes.

Downstream of *ptsR* is a gene for a multiphosphoryl PTS phosphotransferase. The gene, which we have designated *ptsT*, encodes a protein of 808 aa that comprises all three general phosphotransferases of a PTS: IIA (aa 1 to 150), HPr (aa 151 to 251), and EI (aa 251 to 808). The protein resembles the multiphosphoryl phosphotransferase MTP from *Rhodobacter capsulatus* (76), which drives fructose uptake by phosphorylation of the respective FruA PTS permease, in length and do-

TABLE 2. N-terminal sequences of the periplasmic binding proteins of ABC sugar transporters in *M. smegmatis*[a]

Protein	Locus tag	N-terminal sequence (no. of residues)
BglE	msmeg_0505	---------------MTRTRLFRFGSAVASTLTVAALALSACAPGPSGDSGSSPAPTGEVSKDI (49)
AgaE	msmeg_0515	--------------------MIRRWLCLAVVTAVACLLTACGGGSSSSGPVEIAVWHGYQDTE (43)
AbcE	msmeg_0553	----------VTSPAFTRRRALQLLGLAGGAAMLAPALAACGSSGGNSALAADAPVSGRFEGV (53)
RbsB1	msmeg_1374	------------------VNRKRLMLAAGVVALALPMAACTSSKPQADESSETPAAAGEAPA (44)
XylF1	msmeg_1704	-------------------MRKLTWLAALLAALAMAMTLSGCGRSAEGGGGGDGDAKGTVGIAM (45)
AraE	msmeg_1712	--------------------VRKMFAAAIGVVAVAAAVTACGSGKAPGSEGGSAPDGALTLGF (43)
RbsB2	msmeg_3095	------------------VSFAKALSGIALGAAMALSFTGCSVPGDDAAQNAPVVDGALKIGF (45)
AbcE	msmeg_3111	----------MKIPQLRRRRRRPAITAITAMSVAAGLVLSGCAGTGGPANDEATSGVGDVPTDT (54)
AbcE	msmeg_3266	---MMSRESQPGLHRQLSRRNMLAAMGLAGAAAVSLPVLSACGVGGRTNAPNGASEVTGGFDWR (61)
RbsB3	msmeg_3599	------------------MRLGTTAFAIASATALGLGLTACGAGDPAANSDTTRIGVTVYDMS (45)
YphE	msmeg_3999	---------------MPRSLRRRAVRFATLMLVAALAVSGCSRIGENGRIAVVYLNAEGFYAG (48)
RbsB4	msmeg_4172	--------------------MKIRNILPILVCTTCAVAMTACSSSVNNNADPSDTAAPATNVEV (44)
AbcE	msmeg_4658	--------------------MRLSRLVAAAGVGVLMLGASACSSTGGKPDSSGGGDMGAGTADT (44)
SugE	msmeg_5056	--------------------VRARRLCAAAVAAMAAASMVSACGSQTGGIVINYYTPANEEATFK (45)
BxlE	msmeg_5145	VDLKSVDANVVESKADFLPSTSRRAFLAAALSVPLLGALAACGSSGPSRSGGGGGGAPGAASYW (64)
SmoE	msmeg_5574	--------------------MKRLRRLAACIAAAGLTATAGCAGAGTLGATDQTVTIAMVSNSQ (44)
XylE2	msmeg_6020	--------------------VKRTSTLLVTAVVGLGLTLTACGSDSGSNAGSAEGGSGGKIGVI (40)
RbsB5	msmeg_6804	---------------MFRKVTRNTRTVGAALMAGSLVLGMTACGGSGSDGVKVGLITKTDSNPYF (50)

[a] The sequences were aligned according to their signal peptide cleavage site as predicted by SignalP (4). The N-terminal cysteine of the mature protein, which is the catalytic residue of the fatty acid anchor, is marked with an asterisk.

main structure. Hence, PtsT may be involved in the phosphorylation of the adjacent PTS permease PtsG. In addition, PtsT is a homolog of the multiphosphoryl phosphotransferase DhaM of *E. coli*. DhaM undergoes phosphoenolpyruvate-dependent autophosphorylation and then phosphorylates ADP through its enzyme IIA domain. This ATP molecule is used by the dihydroxyacetone kinase DhaKL for substrate phosphorylation (21, 25). Strikingly, homologs of *dhaKL* are, as in *E. coli*, juxtaposed with *ptsT* (*dhaM*). An adjacent permease gene, which we have termed *dhaF*, does not exist in *E. coli*. We predict that PtsT serves as the shuttle for the transfer of phosphate to DhaKL kinase, which in turn phosphorylates dihydroxyacetone molecules that are imported through the MIP family facilitator DhaF.

(ii) Fructose (msmeg_0084 to msmeg_0088). The second PTS locus comprises genes for a fructose-specific PTS composed of EI (*ptsI*), HPr (*ptsH*), and IIABC[Fru] (*fruA*) (Fig. 2B). The locus further consists of a gene coding for a regulator of the DeoR family (51) and a *fruK* gene coding for a protein similar to fructose-1-phosphate kinases (45). The gene order *fruR-fruK-fruA* is the same as what we previously found in *S. coelicolor*, while the *fru* operons of other bacteria are under the control of different regulators (52). The similarities of proteins encoded by genes in these operons are highest between *M. smegmatis* and *S. coelicolor*, with identities of 51% for the DeoR-type regulator, 43% for FruK, and 51% for FruA (IIABC[Fru]).

MIP permeases. Sugar permeases of the MIP family were screened by BLASTP searches with the glycerol facilitator protein sequences of *E. coli* and *S. coelicolor*. We found a protein, msmeg_6758 (GlpF), with identity scores of 42% and 37% to the respective pendants from *E. coli* and *S. coelicolor*. msmeg_6758 is situated in an operon with genes for two glycerol kinases (*glpA1* and *glpA2*) and a glycerol-3-phosphate dehydrogenase (*glpD*) (Fig. 2). A putative regulatory gene (*glpR*) that is homologous to the glycerol operon regulator gene *gylR* of *S. coelicolor* is found upstream (26). Hence, the genes msmeg_6756 to msmeg_6760 are the best candidates to encode proteins for glycerol uptake and metabolism. As described above for PTS, we identified DhaF (msmeg_2124) as being a second MIP family member that may serve as a facilitator for dihydroxyacetone in conjunction with PTS proteins.

MFS permeases. Well-characterized permeases of the MFS family are the xylose symporter XylE of *E. coli* and the glucose-specific symporter GlcP from *S. coelicolor* and the cyanobacterium *Synechocystis* (34, 73, 77). We found four homologs (msmeg_2966, msmeg_4098, msmeg_4182, and msmeg_5559), of which msmeg_4182 exhibited 53% identity to glucose symporters (Fig. 2). msmeg_4182 is surrounded by genes that are not related to sugar metabolism, and there is no putative regulatory gene in the vicinity. Hence, *glcP* might encode a glucose permease that is expressed constitutively in a monocistronic operon. The other three candidates showed very distant similarities to sugar transporters, thus making a substrate prediction impossible.

Growth of *M. smegmatis* on carbohydrates as single carbon sources. According to our genome analysis, *M. smegmatis* possesses uptake systems for a large number of sugars. This is consistent with earlier findings that *M. smegmatis* can grow on many monosaccharides, organic acids, and sugar alcohols (20). In those studies, oxygen consumption of starved bacteria with low endogenous respiration was measured after the addition of single carbon sources to bacteria suspended either in phosphate buffer or in phosphate saline (57). However, in those experiments, *M. smegmatis* may have lacked other elements that may be required to consume the substrate of interest. Therefore, we conducted growth experiments in HB minimal medium, which contains all essential nutrients and trace elements (65), and the carbohydrate of interest as the sole carbon source. *M. smegmatis* readily grew in HB minimal medium with glucose, glycerol, xylose, fructose, ribose, arabinose, or trehalose (Fig. 3 and Table 3). Potential uptake systems were assigned in this study for all of these mono- and disaccharides (see Fig. 5A). These results demonstrated that func-

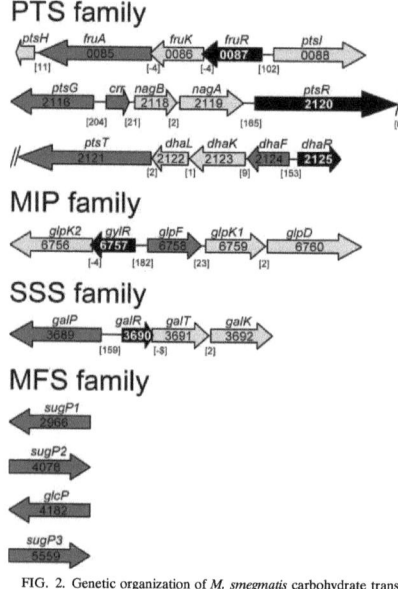

FIG. 2. Genetic organization of *M. smegmatis* carbohydrate transporters of the PTS, MIP, SSS, and MFS protein families. For an explanation of the data, see the legend to Fig. 1.

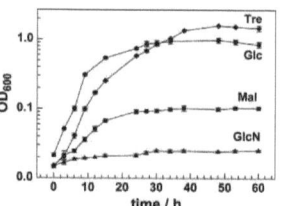

FIG. 3. Growth of *M. smegmatis* in minimal medium with different carbon sources. The growth of *M. smegmatis* mc²155 at 37°C in HB medium containing 1% glucose (circles), 1% trehalose (diamonds), 1% maltose (squares), and 1% GlcNAc (triangles) was measured by determining the OD_{600} of the cultures. The values are the means of three independent experiments. For some data points, the standard deviations were smaller than the symbol size, and therefore, the error bars are invisible.

tional uptake systems exist for these sugars and that these sugars are metabolized by *M. smegmatis*. The generation times ranged from 132 to 174 min, which is similar to growth rates of *M. smegmatis* in Middlebrook 7H9 medium that contains glycerol as the main carbon source (69). The monosaccharides GlcNAc, glucosamine, and galactose; the disaccharides maltose, sucrose, and lactose; and the trisaccharide raffinose were not utilized. This may indicate the lack of uptake and/or lack of metabolic enzymes. Particularly interesting was the growth of *M. smegmatis* on maltose as a sole carbon source. *M. smegmatis* stopped growing at a very early stage at an OD_{600} of 0.1 (Fig. 3). The poor growth of *M. smegmatis* on maltose was not altered when the amount of maltose in the medium was increased from 1% to 10%, indicating that a putative, minor contaminating carbon source did not did not cause the initial residual growth of *M. smegmatis*. Similar results were obtained for galactose, lactose, and sucrose (not shown). The reason for this phenomenon is unknown.

Analysis of the expression of the fructose, glucose, and glycerol import systems. In order to validate the predicted assignment of inner membrane transport systems to a specific sugar, we examined the transcription of selected genes by semiquantitative RT-PCR. First, the transcription of the genes necessary for fructose uptake and utilization were analyzed. The gene msmeg_0086, predicted to encode a fructose-1-phosphate kinase (FruK), showed enhanced mRNA levels in cells grown in the presence of fructose (Fig. 4). Transcription of the adjacent genes *ptsH* and *ptsI*, encoding general components of the PTS, was not enhanced by fructose. Transport experiments with [¹⁴C]fructose were done to examine whether fructose induces the expression of its uptake system. Figure 5A clearly shows that fructose uptake is enhanced in *M. smegmatis* when grown in the presence of 2% fructose. This demonstrates that fructose uptake is inducible in *M. smegmatis*. These results confirm the prediction that msmeg_0085 and msmeg_0086 represent a *fruAK* operon in which *fruA* encodes a fructose-specific enzyme IIA permease, as shown previously for *S. coelicolor* (45). Expression of this operon is most likely controlled by the adjacent DeoR-like regulator that we have thus designated FruR (msmeg_0087).

Two systems were predicted for the uptake of glucose: the symporter GlcP and a glucose-specific PTS. The transcription of *crr* and *ptsT* was induced in cells grown in the presence of glucose (Fig. 4), indicating that their assignments as glucose-

TABLE 3. Growth of *M. smegmatis* on various sugars as a sole carbon source[a]

Sugar	Generation time (min)	Growth[b]
LB	156 ± 24	ND
Glycerol	174 ± 6	+
D-Xylose	192 ± 18	+
D-Ribose	497 ± 36	−/+
L-Arabinose	154 ± 10	+
D-Fructose	174 ± 18	+
D-Glucose	132 ± 12	+
Trehalose	150 ± 12	ND

[a] Cultures of *M. smegmatis* mc²155 were grown on HB minimal medium containing glucose. Cell were washed twice in HB medium, resuspended, and diluted to an OD_{600} of 0.02 in HB medium containing 1% of the respective sugars. The generation times were calculated as described in Materials and Methods. Growth in full medium (LB, lysogeny broth) was used as a control. ND, not determined.

[b] Data reported previously by Izumori et al. (29).

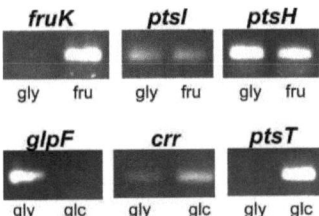

FIG. 4. Expression analysis of selected genes of *M. smegmatis*. The figure shows 1% agarose gels with PCR products from semiquantitative RT-PCR experiments. For each gene, samples were taken periodically along the PCR. The depicted bands show products from the same cycle for each gene, when the amplification was in the linear range. RT-PCR experiments without a reverse transcriptase reaction delivered no signal (negative control). Data were obtained from two biological replicates. gly, glycerol; fru, fructose; glc, glucose.

specific PTS enzyme IIA and tripartite PTS phosphotransferase are correct. Uptake of [^{14}C]glucose by *M. smegmatis* was not altered when cells were grown in the absence or presence of 2% glucose (Fig. 5B). One possibility to explain this result is that *glcP* is expressed constitutively. However, the *glcP* mRNA was not detected in RT-PCR experiments for unknown reasons, in contrast to mRNA of control genes such as *sigA* (not shown). An alternative explanation is that the rate-limiting step for glucose uptake is diffusion across the outer membrane of *M. smegmatis*. Indeed, porin mutants of *M. smegmatis* show a significantly impaired glucose uptake (67, 69). To date, it is unknown whether porin-mediated diffusion of glucose is the rate-limiting step in wild-type *M. smegmatis*. The fact that uninduced glucose uptake is as fast as the induced uptake of fructose (Fig. 5) argues in favor of a constitutively expressed inner membrane transporter for glucose. Further experiments with mutants lacking either GlcP or PtsG are required to examine whether glucose transport is regulated in *M. smegmatis*.

The number of transcripts of the predicted glycerol facilitator gene *glpF* (msmeg_6758) was increased in cells grown in the presence of 0.4% glycerol compared to that in cells grown in the presence of 0.4% glucose (Fig. 4). Furthermore, the uptake of [^{14}C]glycerol was slightly increased in *M. smegmatis* cells grown in the presence of 0.4% glycerol compared to that in cells grown in the presence of 0.4% glucose (Fig. 5C). These results confirmed the annotation of the msmeg_6758 gene as *glpF*.

Carbohydrate transporters of *M. tuberculosis* H37Rv. We evaluated the complement of the putative carbohydrate uptake systems in the slow-growing pathogenic strain *M. tuberculosis* H37Rv (13). Our analysis resulted in the identification of four ABC-type systems and one permease of the MFS class (Fig. 6B). Some of these ABC transporters were described previously in an in silico analysis of the *M. tuberculosis* genome in a more global context (9, 15). It is striking that *M. tuberculosis* is poorly equipped with carbohydrate transport systems in comparison to *M. smegmatis* mc^{2}155 (Fig. 6). Two of the operons, the *lpgY sugABC* and the *uspABC* operons, are highly conserved between the two species. The proteins of the ABC^{Sug}

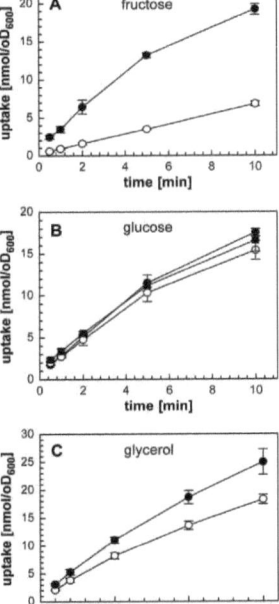

FIG. 5. Kinetics and inducibility of fructose, glucose, and glycerol uptake by *M. smegmatis*. (A) *M. smegmatis* mc^{2}155 was grown in Middlebrook 7H9 medium in the presence of 2% glycerol (open circles) or 2% fructose (closed circles). Accumulation of [^{14}C]fructose was measured at 25°C at a final fructose concentration of 20 μM. (B) *M. smegmatis* mc^{2}155 was grown in Middlebrook 7H9 medium in the presence of 2% glycerol (open circles), 2% glucose (closed circles), and 2% glucose plus 2% glucose (closed squares). Accumulation of [^{14}C]glucose was measured at 25°C at a final glucose concentration of 20 μM. (C) *M. smegmatis* mc^{2}155 was grown in HB medium containing 0.4% glucose (open circles) and 0.4% glycerol (closed circles) to an OD$_{600}$ of 0.9. Accumulation of [^{14}C]glycerol was measured at 37°C at a final concentration of 100 μM glycerol. For A and B, the standard deviations are indicated by error bars and represent the means of three independent experiments. For C, samples were taken in duplicates from two biological replicates.

and of the ABC^{Usp} systems share between 62% and 80% similar amino acids, compared to only 25 to 30% similar amino acids for the UgpABCE and Rv2038c-Rv2041c systems. The similarities of all four ABC systems to known transporters outside the genus *Mycobacterium* is so low (<25%) that substrates of these transporters cannot be predicted.

The SugI porter of the MFS class shows distant sequence similarity to the glucose permease GlcP (28%) of *S. coelicolor*

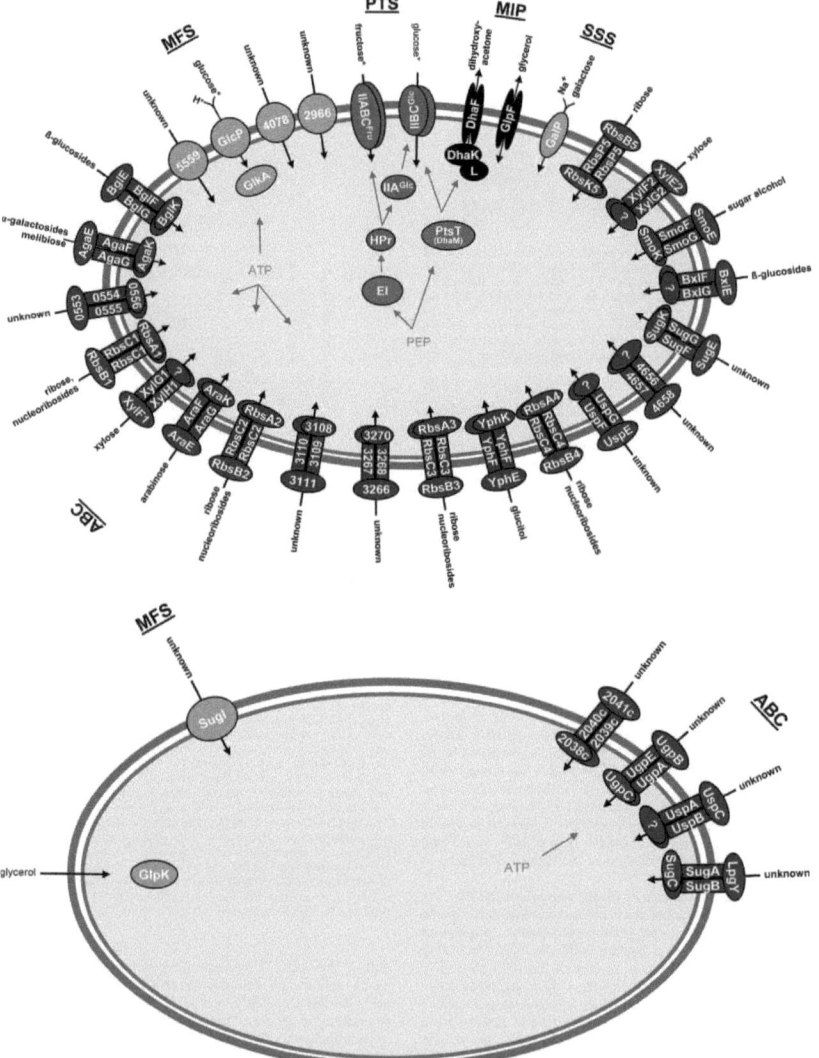

FIG. 6. Sugar transport systems of *M. smegmatis* and *M. tuberculosis*. Shown are the permeases of the ABC, PTS, MIP, MFS, and SSS families. The derived putative substrates are inferred from in silico analyses in combination with experimental data.

and to the galactose (GalP) (24%) and arabinose (AraE) (24%) transporters of *E. coli*. Thus, the system is likely to transport a monosaccharide.

Glycerol is used as the standard carbon source to grow *M. tuberculosis*. We did not detect any putative uptake system for this carbohydrate. Since *M. tuberculosis* grows with a generation time of 24 h, and it has been shown that glycerol can directly diffuse through lipid membranes both in vitro (50) and in vivo (22), it is conceivable that the rate of glycerol intake by passive diffusion may be sufficient for growth. Incoming glycerol will then be converted by glycerol kinase (GlpK) into glycerol-3-phosphate to enter the route of central carbon metabolism (Fig. 6B). *M. tuberculosis* has one putative glycerol kinase that shows a high similarity to the two glycerol kinases of *M. smegmatis* (77% protein identity for msmeg_6759 and 57% for msmeg_6756) and to the two glycerol kinases from *S. coelicolor* SCO0509 (75%) and SCO1660 (59%).

DISCUSSION

Identification of carbohydrate transporters in *M. smegmatis*. It has been widely documented that *M. smegmatis* can grow on many carbon sources such as polyols, pentoses, and hexoses (20, 23, 28). In this study, we identified multiple inner membrane transport systems for all three of these classes of carbohydrates by bioinformatic analysis (Table 1). This provides the molecular basis for the adaptability of *M. smegmatis* to different environments in the soil and water. Often, our integrated bioinformatic approach enabled us to propose a specific substrate for particular uptake proteins (Table 1). Since the specificity of transport proteins can be altered by the modification of a few residues, the suggested substrates rather represent a hypothesis for experiments such as transport measurements with gene deletion mutants or analysis of the induction of gene expression. Analysis of the induction of both the transport activity and transcription of genes confirmed the substrate predictions for a fructose- and glucose-specific PTS as well as for the predicted glycerol operon. Since glucose uptake was constitutive, at least one more system for glucose transport must occur in *M. smegmatis*. This was predicted to be GlcP. Indeed, cloning and heterologous expression of *glcP* in *E. coli* revealed that it is a glucose-specific permease (data not shown).

Global control of carbon metabolism in *M. smegmatis* and *M. tuberculosis*. The discovery of homologs of all components of a PTS in *M. smegmatis* contradicts a previous report that did not find biochemical evidence for the existence of a PTS (59) and many repeating statements (14, 15, 54). Components of the PTS play a key role in the global control of sugar metabolism to achieve the hierarchical utilization of carbon sources in bacteria (12), where two different mechanisms have evolved. In *E. coli* and other closely related gram-negative bacteria, the enzyme IIA^{Glc} is dephosphorylated mainly under repressing conditions and mediates inducer exclusion. Under nonrepressing conditions, phosphorylated IIA^{Glc} stimulates cyclic AMP (cAMP) synthesis and thereby triggers the activation of catabolite-repressed genes by a global regulator, the cAMP-dependent catabolite activator protein. In low-G+C-content gram-positive bacteria, HPr is a central switch of carbon catabolite repression. Under repressing conditions, HPr is phosphorylated mainly at serine 46 by a unique HPr kinase/phosphatase mediating inducer exclusion and carbon catabolite repression/activation (12, 37, 55). Under nonrepressing conditions, HPr is phosphorylated at histidine 15 and activates PTS-dependent sugar transport, glycerol kinase, and substrate-specific regulators. The apparent absence of a protein in *M. smegmatis* similar to the HPr kinase/phosphatase argues against the mechanism found in low-G+C-content gram-positive bacteria. On the other hand, *M. smegmatis* apparently does not produce proteins with significant sequence similarities to the cAMP receptor protein (CRP) (catabolite activator protein) of *E. coli*, which is crucial for carbon catabolite repression in gram-negative bacteria.

By contrast, the coordination of the few operons involved in the uptake and degradation of carbohydrates by *M. tuberculosis* may not require a global control mechanism, as suggested by the lack of PTS homologs. Alternatively, a completely different mechanism for the global control of carbon metabolism may have evolved in *M. tuberculosis* to adapt to its specific environment inside the phagosome of human macrophages. Indeed, *M. tuberculosis* has eight orthologs of CRP-like transcriptional regulators (35), one of which, Rv3676, was experimentally described (2). Furthermore, the large number and the different subcellular localization of the 15 putative nucleotide cyclases in *M. tuberculosis* imply that this organism may have the ability to sense and respond to many intracellular and extracellular signals through a second messenger system based on cyclic nucleotide monophosphates (35). This is in strong contrast to *E. coli* and other gram-negative bacteria, which have only one CRP and one adenylate cyclase. CRP homologs have been identified in streptomycetes (18), where the regulator plays a role in germination, and in corynebacteria, where CRPs have been associated with global carbon regulation (39). Although the potential mechanisms of global control of carbon metabolism in both *M. smegmatis* and *M. tuberculosis* are not evident from the bioinformatic analysis of their genomes, these findings provide hypotheses for further experiments.

Utilization of galactose by *M. smegmatis* and *M. tuberculosis*. *M. smegmatis* did not grow on D-galactose as a sole carbon source. This result is in agreement with previous reports (28, 29). Surprisingly, our analysis identified the genes msmeg_3689 to msmeg_3692 as being an operon encoding, among others, a putative galactose transport protein of the SSS with more than 50% amino acid identity to the putative galactose transporter GalP of *S. coelicolor* (6). Furthermore, it was shown that D-galactose is taken up by *M. smegmatis* and that this transport activity is inducible by D-galactose (28). However, the putative *gal* operon of *M. smegmatis* contains only two of three essential genes (*galKTE*) of the Leloir pathway, which is used by *E. coli* and many other bacteria to grow on galactose (24). The first reaction in this pathway is catalyzed by galactokinase, which phosphorylates free galactose to galactose-1-phosphate. In the next steps, galactose-1-phosphate uridylyltransferase transfers the UDP residue from UDP-glucose to galactose-1-phosphate, and UDP-galactose-4-epimerase catalyzes the reversible conversions of UDP-galactose and UDP-glucose. The *galE* gene encoding the epimerase is missing. These results appear to be counterintuitive: why should *M. smegmatis* take up a sugar, modify it, and not use it as a carbon or energy source? One explanation may be provided by the fact that D-galactose is a major constituent of the cell wall of mycobacteria. The arabinogalactan polymer is composed of D-arabinose and D-galactose, both in their furanose ring form, and is an essential part of the mycobacterial cell wall by serving as an anchor for the covalent attachment of the peptidoglycan and the mycolic acids (10). Ethambutol is a potent tuberculosis drug and acts by inhibiting the synthesis of arabinogalactan (71). Based on these observations, we suggest the following sequence of reactions: D-galactose is taken up by *M. smegmatis* via GalP (msmeg_3689), phosphorylated in the cytoplasm by the galactokinase GalK (msmeg_3692), and uridinylated by the galactose-1-phosphate uridyltransferase GalT (msmeg_3691) to yield UDP-galactose. The ring contraction of UDP-galactopyranose to UDP-galactofuranose is catalyzed by the essential enzyme UDP-galactopyranose mutase Glf (46). UDP-galactofuranose is then likely to be transported across the cytoplasm membrane via intermediate binding to decaprenyl phosphate as in other bacteria to be available for the synthesis of arabinogalactan (16). It should be noted that no homologs of the putative galactose transporter GalP of *M. smegmatis* (Table 1) were found in *M. tuberculosis*. This suggests that *M. tuberculosis* has no access to D-galactose in its natural environment, the phagosome of macrophages, and instead synthesizes D-galactose from other sugars. Indeed, in addition to *galK* and *galT*, the genome of *M. tuberculosis* contains three *galE* genes (Rv3634c, Rv0501, and Rv0536), which are probably used to convert glucose to galactose for biosynthesis purposes. However, it is unclear how *M. smegmatis* synthesizes galactose in the absence of this sugar and GalE. *M. smegmatis* may either contain an undetected enzyme with glucose-galactose epimerase activity or use an alternative biosynthetic route.

Utilization of disaccharides by *M. smegmatis* and *M. tuberculosis*. *M. smegmatis* did not grow on lactose, maltose, and sucrose as a sole carbon source. Franke and Schillinger previously obtained the same result for lactose and maltose but observed respiration of *M. smegmatis* in the presence of sucrose (23). According to our bioinformatic analysis, *M. smegmatis* has at least three inner membrane transport systems with significant similarities to other bacterial disaccharide transporters (Table 1). However, the substrate specificities of the transporters encoded within the loci msmeg_0501 to msmeg_0508 and msmeg_0509 to msmeg_0518 are not known. Growth of bacteria on disaccharides as sole carbon sources requires enzymes that cleave the disaccharide and release the monosaccharides for further metabolization. The absence of proteins similar to known bacterial β-D-galactosidases (LacZ of *E. coli*, BgaB of *Bacillus circulans*, MbgA of *Bacillus megaterium*, and LacA of *S. coelicolor*) provides a molecular explanation for the inability of *M. smegmatis* to utilize lactose as a sole carbon source. By contrast, *M. smegmatis* has six homologs (msmeg_3184, msmeg_3576, msmeg_4916, msmeg_4917, msmeg_4696, and msmeg_6515) of MalL of *B. subtilis*, which hydrolyzes maltose, longer maltodextrines up to maltohexaose, isomaltose, and sucrose (64), and of the cytoplasmic trehalase TreC of *E. coli*, which cleaves trehalose-6-phosphate (58). It is conceivable that these enzymes are used in trehalose metabolism, considering the unusual importance of trehalose in mycobacteria (42, 74) and the observation that trehalose was the only disaccharide that was used by *M. smegmatis* as a sole carbon source. However, it cannot be excluded that some of the enzymes with similarities to TreC and MalF have roles in pathways distinct from trehalose metabolism.

The SugABC sugar transport system was shown to be essential for the virulence of *M. tuberculosis* in mice (61). Previously, it was suggested that this permease may transport maltose or maltodextrins (8, 9). However, the similarities of both ABC^{Sug} and the corresponding substrate binding protein LpgY to the maltose transporters and periplasmic maltose binding proteins MalE of *E. coli* and *S. coelicolor* are very low (<25%). Thus, it is questionable whether maltose is the substrate of ABC^{Sug}. These doubts are supported by the fact that neither *M. smegmatis*, which has a highly similar ABC^{Sug} system, nor *M. tuberculosis* (20) grows on maltose as a sole carbon source. It has to be noted that similar uncertainties exist for the substrate specificities of the four other carbohydrate uptake systems of *M. tuberculosis*, including the ABC^{Usp} transporter, which was proposed to transport *sn*-glycerol-3-phosphate based on low protein similarities (9, 15).

The analysis of the carbohydrate uptake proteins in the genomes of *M. smegmatis* and *M. tuberculosis* provides the molecular basis for the very early phenotypic observations that saprophytic mycobacteria have a much broader spectrum of substrates, which they can use as sole carbon and energy sources (20). It is striking that the genome of *M. tuberculosis* has only five recognizable permeases for carbohydrate uptake. This suggests that the phagosome does not provide an environment rich in diverse sugars. Hence, an experimental analysis of the substrate specificity of the inner membrane carbohydrate transporters of *M. tuberculosis* is likely to reveal the carbon sources available in the phagosome of human macrophages.

ACKNOWLEDGMENTS

This work was supported by grant AI06432 from the National Institutes of Health to M.N. and by grants SFB473 and Graduiertenkolleg GK805 of the Deutsche Forschungsgemeinschaft.

Sequencing of *M. smegmatis* mc^{2}155 was accomplished by TIGR with support from National Institute of Allergy and Infectious Diseases (NIAID). We thank Natalie Wood, Flavia Pimentel-Schmitt, Hildegard Stork, and Ying Wang for technical assistance.

REFERENCES

1. **Alvarez-Anorve, L. I., M. L. Calcagno, and J. Plumbridge.** 2005. Why does *Escherichia coli* grow more slowly on glucosamine than on *N*-acetylglucosamine? Effects of enzyme levels and allosteric activation of GlcN6P deaminase (NagB) on growth rates. J. Bacteriol. **187:**2974–2982.
2. **Bai, G., L. A. McCue, and K. A. McDonough.** 2005. Characterization of *Mycobacterium tuberculosis* Rv3676 (CRPMt), a cyclic AMP receptor protein-like DNA binding protein. J. Bacteriol. **187:**7795–7804.
3. **Bell, A. W., S. D. Buckel, J. M. Groarke, J. N. Hope, D. H. Kingsley, and M. A. Hermodson.** 1986. The nucleotide sequences of the *rbsD*, *rbsA*, and *rbsC* genes of *Escherichia coli* K12. J. Biol. Chem. **261:**7652–7658.
4. **Bendtsen, J. D., H. Nielsen, G. von Heijne, and S. Brunak.** 2004. Improved prediction of signal peptides: SignalP 3.0. J. Mol. Biol. **340:**783–795.
5. **Bentley, S. D., K. F. Chater, A. M. Cerdeno-Tarraga, G. L. Challis, N. R. Thomson, K. D. James, D. E. Harris, M. A. Quail, H. Kieser, D. Harper, A. Bateman, S. Brown, G. Chandra, C. W. Chen, M. Collins, A. Cronin, A. Fraser, A. Goble, J. Hidalgo, T. Hornsby, S. Howarth, C. H. Huang, T. Kieser, L. Larke, L. Murphy, K. Oliver, S. O'Neil, E. Rabbinowitsch, M. A. Rajandream, K. Rutherford, S. Rutter, K. Seeger, D. Saunders, S. Sharp, R. Squares, S. Squares, K. Taylor, T. Warren, A. Wietzorrek, J. Woodward, B. G. Barrell, J. Parkhill, and D. A. Hopwood.** 2002. Complete genome sequence of the model actinomycete *Streptomyces coelicolor* A3(2). Nature **417:**141–147.
6. **Bertram, R., M. Schlicht, K. Mahr, H. Nothaft, M. H. Saier, Jr., and F. Titgemeyer.** 2004. In silico and transcriptional analysis of carbohydrate uptake systems of *Streptomyces coelicolor* A3(2). J. Bacteriol. **186:**1362–1373.
7. **Blattner, F. R., G. Plunkett III, C. A. Bloch, N. T. Perna, V. Burland, M.**

Riley, J. Collado-Vides, J. D. Glasner, C. K. Rode, G. F. Mayhew, J. Gregor, N. W. Davis, H. A. Kirkpatrick, M. A. Goeden, D. J. Rose, B. Mau, and Y. Shao. 1997. The complete genome sequence of *Escherichia coli* K-12. Science **277:**1453–1474.
8. Borich, S. M., A. Murray, and E. Gormley. 2000. Genomic arrangement of a putative operon involved in maltose transport in the *Mycobacterium tuberculosis* complex and *Mycobacterium leprae*. Microbios **102:**7–15.
9. Braibant, M., P. Gilot, and J. Content. 2000. The ATP binding cassette (ABC) transport systems of *Mycobacterium tuberculosis*. FEMS Microbiol. Rev. **24:**449–467.
10. Brennan, P. J., and H. Nikaido. 1995. The envelope of mycobacteria. Annu. Rev. Biochem. **64:**29–63.
11. Brinkkötter, A., H. Kloss, C. Alpert, and J. W. Lengeler. 2000. Pathways for the utilization of N-acetyl-galactosamine and galactosamine in *Escherichia coli*. Mol. Microbiol. **37:**125–135.
12. Brückner, R., and F. Titgemeyer. 2002. Carbon catabolite repression in bacteria: choice of the carbon source and autoregulatory limitation of sugar utilization. FEMS Microbiol. Lett. **209:**141–148.
13. Cole, S. T., R. Brosch, J. Parkhill, T. Garnier, C. Churcher, D. Harris, S. V. Gordon, K. Eiglmeier, S. Gas, C. E. Barry III, F. Tekaia, K. Badcock, D. Basham, D. Brown, T. Chillingworth, R. Connor, R. Davies, K. Devlin, T. Feltwell, S. Gentles, N. Hamlin, S. Holroyd, T. Hornsby, K. Jagels, A. Krogh, J. McLean, S. Moule, L. Murphy, K. Oliver, J. Osborne, M. A. Quail, M. A. Rajandream, J. Rogers, S. Rutter, K. Seeger, J. Skelton, R. Squares, S. Squares, J. E. Sulston, K. Taylor, S. Whitehead, and B. G. Barrell. 1998. Deciphering the biology of *Mycobacterium tuberculosis* from the complete genome sequence. Nature **393:**537–544.
14. Connell, N. D., and H. Nikaido. 1994. Membrane permeability and transport in *Mycobacterium tuberculosis*, p. 333–349. In B. R. Bloom (ed.), Tuberculosis: pathogenesis, protection, and control. American Society for Microbiology, Washington, DC.
15. Content, J., M. Braibant, N. Connell, and J. A. Ainsa. 2005. Transport processes, p. 379–404. In S. T. Cole, D. N. Eisenach, D. N. McMurray, and W. R. Jacobs (ed.), Tuberculosis and the tubercle bacillus, ASM Press, Washington, DC.
16. Crick, D. C., S. Mahapatra, and P. J. Brennan. 2001. Biosynthesis of the arabinogalactan-peptidoglycan complex of *Mycobacterium tuberculosis*. Glycobiology **11:**107R–118R.
17. Daley, D. O., M. Rapp, E. Granseth, K. Melen, D. Drew, and G. von Heijne. 2005. Global topology analysis of the *Escherichia coli* inner membrane proteome. Science **308:**1321–1323.
18. Derouaux, A., S. Halici, H. Nothaft, T. Neutelings, G. Moutzourelis, J. Dusart, F. Titgemeyer, and S. Rigali. 2004. Deletion of a cyclic AMP receptor protein homologue diminishes germination and affects morphological development of *Streptomyces coelicolor*. J. Bacteriol. **186:**1893–1897.
19. Duplay, P., H. Bedouelle, A. Fowler, I. Zabin, W. Saurin, and M. Hofnung. 1984. Sequences of the malE gene and of its product, the maltose-binding protein of *Escherichia coli* K12. J. Biol. Chem. **259:**10606–10613.
20. Edson, N. L. 1951. The intermediary metabolism of the mycobacteria. Bacteriol. Rev. **15:**147–182.
21. Erni, B., C. Siebold, S. Christen, A. Srinivas, A. Oberholzer, and U. Baumann. 2006. Small substrate, big surprise: fold, function and phylogeny of dihydroxyacetone kinases. Cell. Mol. Life Sci. **63:**890–900.
22. Eze, M. O., and R. N. McElhaney. 1981. The effect of alterations in the fluidity and phase state of the membrane lipids on the passive permeation and facilitated diffusion of glycerol in *Escherichia coli*. J. Gen. Microbiol. **124:**299–307.
23. Franke, W., and A. Schillinger. 1944. Zum Stoffwechsel der saeurefesten Bakterien I. Orientierende aerobe Reihenversuche. Biochem. Zeitung. **319:**313–334.
24. Frey, P. A. 1996. The Leloir pathway: a mechanistic imperative for three enzymes to change the stereochemical configuration of a single carbon in galactose. FASEB J. **10:**461–470.
25. Gutknecht, R., R. Beutler, L. F. Garcia-Alles, U. Baumann, and B. Erni. 2001. The dihydroxyacetone kinase of *Escherichia coli* utilizes a phosphoprotein instead of ATP as phosphoryl donor. EMBO J. **20:**2480–2486.
26. Hindle, Z., and C. P. Smith. 1994. Substrate induction and catabolite repression of the *Streptomyces coelicolor* glycerol operon are mediated through the GylR protein. Mol. Microbiol. **12:**737–745.
27. Hurtubise, Y., F. Shareck, D. Kluepfel, and R. Morosoli. 1995. A cellulase/xylanase-negative mutant of Streptomyces lividans 1326 defective in cellobiose and xylobiose uptake is mutated in a gene encoding a protein homologous to ATP-binding proteins. Mol. Microbiol. **17:**367–377.
28. Izumori, K., Y. Ueda, and K. Yamanaka. 1978. Pentose metabolism in *Mycobacterium smegmatis*: comparison of L-arabinose isomerases induced by L-arabinose and D-galactose. J. Bacteriol. **133:**413–414.
29. Izumori, K., K. Yamanaka, and D. Elbein. 1976. Pentose metabolism in *Mycobacterium smegmatis*: specificity of induction of pentose isomerases. J. Bacteriol. **128:**587–591.
30. Kamionka, A., S. Parche, H. Nothaft, J. Siepelmeyer, K. Jahreis, and F. Titgemeyer. 2002. The phosphotransferase system of *Streptomyces coelicolor*. Eur. J. Biochem. **269:**2143–2150.

31. Kawaguchi, H., A. A. Vertes, S. Okino, M. Inui, and H. Yukawa. 2006. Engineering of a xylose metabolic pathway in *Corynebacterium glutamicum*. Appl. Environ. Microbiol. **72:**3418–3428.
32. Koch, R. 1882. Die Aetiologie der Tuberculose. Berl. Klin. Wochenschr. **19:**221–230.
33. Lawlis, V. B., M. S. Dennis, E. Y. Chen, D. H. Smith, and D. J. Henner. 1984. Cloning and sequencing of the xylose isomerase and xylulose kinase genes of *Escherichia coli*. Appl. Environ. Microbiol. **47:**15–21.
34. Maiden, M. C., E. O. Davis, S. A. Baldwin, D. C. Moore, and P. J. Henderson. 1987. Mammalian and bacterial sugar transport proteins are homologous. Nature **325:**641–643.
35. McCue, L. A., K. A. McDonough, and C. E. Lawrence. 2000. Functional classification of cNMP-binding proteins and nucleotide cyclases with implications for novel regulatory pathways in *Mycobacterium tuberculosis*. Genome Res. **10:**204–219.
36. McKinney, J. D., K. Honer zu Bentrup, E. J. Munoz-Elias, A. Miczak, B. Chen, W. T. Chan, D. Swenson, J. C. Sacchettini, W. R. Jacobs, Jr., and D. G. Russell. 2000. Persistence of *Mycobacterium tuberculosis* in macrophages and mice requires the glyoxylate shunt enzyme isocitrate lyase. Nature **406:**735–738.
37. Monedero, V., A. Maze, G. Boel, M. Zuniga, S. Beaufils, A. Hartke, and J. Deutscher. 2007. The phosphotransferase system of *Lactobacillus casei*: regulation of carbon metabolism and connection to cold shock response. J. Mol. Microbiol. Biotechnol. **12:**20–32.
38. Montero-Moran, G. M., S. Lara-Gonzalez, L. I. Alvarez-Anorve, J. A. Plumbridge, and M. L. Calcagno. 2001. On the multiple functional roles of the active site histidine in catalysis and allosteric regulation of *Escherichia coli* glucosamine 6-phosphate deaminase. Biochemistry **40:**10187–10196.
39. Moon, M. W., S. Y. Park, S. K. Choi, and J. K. Lee. 2007. The phosphotransferase system of *Corynebacterium glutamicum*: features of sugar transport and carbon regulation. J. Mol. Microbiol. Biotechnol. **12:**43–50.
40. Mota, L. J., P. Tavares, and I. Sa-Nogueira. 1999. Mode of action of AraR, the key regulator of L-arabinose metabolism in *Bacillus subtilis*. Mol. Microbiol. **33:**476–489.
41. Munoz-Elias, E. J., and J. D. McKinney. 2005. *Mycobacterium tuberculosis* isocitrate lyases 1 and 2 are jointly required for in vivo growth and virulence. Nat. Med. **11:**638–644.
42. Murphy, H. N., G. R. Stewart, V. V. Mischenko, A. S. Apt, R. Harris, M. S. McAlister, C. P. Driscoll, D. B. Young, and B. D. Robertson. 2005. The OtsAB pathway is essential for trehalose biosynthesis in *Mycobacterium tuberculosis*. J. Biol. Chem. **280:**14524–14529.
43. Nobelmann, B., and J. W. Lengeler. 1996. Molecular analysis of the gat genes from *Escherichia coli* and of their roles in galactitol transport and metabolism. J. Bacteriol. **178:**6790–6795.
44. Nothaft, H., D. Dresel, A. Willimek, K. Mahr, M. Niederweis, and F. Titgemeyer. 2003. The phosphotransferase system of *Streptomyces coelicolor* is biased for N-acetylglucosamine metabolism. J. Bacteriol. **185:**7019–7023.
45. Nothaft, H., S. Parche, A. Kamionka, and F. Titgemeyer. 2003. In vivo analysis of HPr reveals a fructose-specific phosphotransferase system that confers high-affinity uptake in *Streptomyces coelicolor*. J. Bacteriol. **185:**929–937.
46. Pan, F., M. Jackson, Y. Ma, and M. McNeil. 2001. Cell wall core galactofuran synthesis is essential for growth of mycobacteria. J. Bacteriol. **183:**3991–3998.
47. Pao, S. S., I. T. Paulsen, and M. H. Saier, Jr. 1998. Major facilitator superfamily. Microbiol. Mol. Biol. Rev. **62:**1–34.
48. Parche, S., H. Nothaft, A. Kamionka, and F. Titgemeyer. 2000. Sugar uptake and utilisation in *Streptomyces coelicolor*: a PTS view to the genome. Antonie Leeuwenhoek **78:**243–251.
49. Park, J. H., and M. H. Saier, Jr. 1996. Phylogenetic characterization of the MIP family of transmembrane channel proteins. J. Membr. Biol. **153:**171–180.
50. Paula, S., A. G. Volkov, A. N. V. Hoek, T. H. Haines, and D. W. Deamer. 1996. Permeation of protons, potassium ions, and small polar molecules through phospholipid bilayers as a function of membrane thickness. Biophys. J. **70:**339–348.
51. Perez-Rueda, E., and J. Collado-Vides. 2000. The repertoire of DNA-binding transcriptional regulators in *Escherichia coli* K-12. Nucleic Acids Res. **28:**1838–1847.
52. Postma, P. W., J. W. Lengeler, and G. R. Jacobson. 1993. Phosphoenolpyruvate:carbohydrate phosphotransferase systems of bacteria. Microbiol. Rev. **57:**543–594.
53. Ramakrishnan, T., P. S. Murthy, and K. P. Gopinathan. 1972. Intermediary metabolism of mycobacteria. Bacteriol. Rev. **36:**65–108.
54. Ratledge, C., and J. Stanford. 1982. Nutrition, growth, and metabolism, p. 186–271. In C. Ratledge and J. Stanford (ed.), The biology of mycobacteria. Academic Press Inc. Ltd., London, United Kingdom.
55. Reizer, J., C. Hoischen, F. Titgemeyer, C. Rivolta, R. Rabus, J. Stülke, D. Karamata, M. H. Saier, Jr., and W. Hillen. 1998. A novel protein kinase that controls catabolite repression in bacteria. Mol. Microbiol. **27:**1157–1169.
56. Reizer, J., A. Reizer, and M. H. Saier, Jr. 1994. A functional superfamily of sodium/solute symporters. Biochim. Biophys. Acta **1197:**133–166.
57. Richardson, H. B., E. Shorr, and R. O. Loebel. 1931. Comparative studies in

the respiratory metabolism of various acid-fast bacilli. Trans. Nat. Tuberc. Assoc. **27:**205–210.
58. **Rimmele, M., and W. Boos.** 1994. Trehalose-6-phosphate hydrolase of *Escherichia coli.* J. Bacteriol. **176:**5654–5664.
59. **Romano, A. H., S. J. Eberhard, S. L. Dingle, and T. D. McDowell.** 1970. Distribution of the phosphoenolpyruvate:glucose phosphotransferase system in bacteria. J. Bacteriol. **104:**808–813.
60. **Sa-Nogueira, I., T. V. Nogueira, S. Soares, and H. de Lencastre.** 1997. The *Bacillus subtilis* L-arabinose (ara) operon: nucleotide sequence, genetic organization and expression. Microbiology **143:**957–969.
61. **Sassetti, C. M., and E. J. Rubin.** 2003. Genetic requirements for mycobacterial survival during infection. Proc. Natl. Acad. Sci. USA **100:**12989–12994.
62. **Schlösser, A., J. Jantos, K. Hackmann, and H. Schrempf.** 1999. Characterization of the binding protein-dependent cellobiose and cellotriose transport system of the cellulose degrader *Streptomyces reticuli.* Appl. Environ. Microbiol. **65:**2636–2643.
63. **Schlösser, A., and H. Schrempf.** 1996. A lipid-anchored binding protein is a component of an ATP-dependent cellobiose/cellotriose-transport system from the cellulose degrader *Streptomyces reticuli.* Eur. J. Biochem. **242:**332–338.
64. **Schönert, S., T. Buder, and M. K. Dahl.** 1999. Properties of maltose-inducible alpha-glucosidase MalL (sucrase-isomaltase-maltase) in *Bacillus subtilis*: evidence for its contribution to maltodextrin utilization. Res. Microbiol. **150:**167–177.
65. **Smeulders, M. J., J. Keer, R. A. Speight, and H. D. Williams.** 1999. Adaptation of *Mycobacterium smegmatis* to stationary phase. J. Bacteriol. **181:**270–283.
66. **Soerensen, K. I., and B. Hove-Jensen.** 1996. Ribose catabolism of *Escherichia coli*: characterization of the *rpiB* gene encoding ribose phosphate isomerase B and of the *rpiR* gene, which is involved in regulation of *rpiB* expression. J. Bacteriol. **178:**1003–1011.
67. **Stahl, C., S. Kubetzko, I. Kaps, S. Seeber, H. Engelhardt, and M. Niederweis.** 2001. MspA provides the main hydrophilic pathway through the cell wall of *Mycobacterium smegmatis.* Mol. Microbiol. **40:**451–464.
68. **Stephan, J., J. G. Bail, F. Titgemeyer, and M. Niederweis.** 2004. DNA-free RNA preparations from mycobacteria. BMC Microbiol. **4:**45.
69. **Stephan, J., J. Bender, F. Wolschendorf, C. Hoffmann, E. Roth, C. Mailländer, H. Engelhardt, and M. Niederweis.** 2005. The growth rate of *Mycobacterium smegmatis* depends on sufficient porin-mediated influx of nutrients. Mol. Microbiol. **58:**714–730.
70. **Sumiya, M., E. O. Davis, L. C. Packman, T. P. McDonald, and P. J. Henderson.** 1995. Molecular genetics of a receptor protein for D-xylose, encoded by the gene *xylF*, in *Escherichia coli.* Receptors Channels **3:**117–128.
71. **Takayama, K., and J. O. Kilburn.** 1989. Inhibition of synthesis of arabinogalactan by ethambutol in *Mycobacterium smegmatis.* Antimicrob. Agents Chemother. **33:**1493–1499.
72. **Tobisch, S., P. Glaser, S. Krüger, and M. Hecker.** 1997. Identification and characterization of a new β-glucoside utilization system in *Bacillus subtilis.* J. Bacteriol. **179:**496–506.
73. **van Wezel, G. P., K. Mahr, M. Konig, B. A. Traag, E. F. Pimentel-Schmitt, A. Willimek, and F. Titgemeyer.** 2005. GlcP constitutes the major glucose uptake system of *Streptomyces coelicolor* A3(2). Mol. Microbiol. **55:**624–636.
74. **Woodruff, P. J., B. L. Carlson, B. Siridechadilok, M. R. Pratt, R. H. Senaratne, J. D. Mougous, L. W. Riley, S. J. Williams, and C. R. Bertozzi.** 2004. Trehalose is required for growth of *Mycobacterium smegmatis.* J. Biol. Chem. **279:**28835–28843.
75. **Woodson, K., and K. M. Devine.** 1994. Analysis of a ribose transport operon from *Bacillus subtilis.* Microbiology **140:**1829–1838.
76. **Wu, L. F., J. M. Tomich, and M. H. Saier, Jr.** 1990. Structure and evolution of a multidomain multiphosphoryl transfer protein. Nucleotide sequence of the *fruB*(HI) gene in *Rhodobacter capsulatus* and comparisons with homologous genes from other organisms. J. Mol. Biol. **213:**687–703.
77. **Zhang, C. C., M. C. Durand, R. Jeanjean, and F. Joset.** 1989. Molecular and genetical analysis of the fructose-glucose transport system in the cyanobacterium *Synechocystis* PCC6803. Mol. Microbiol. **3:**1221–1229.

Research Article

J Mol Microbiol Biotechnol
DOI: 10.1159/000119546

Published online: March 3, 2008

Identification of a Glucose Permease from *Mycobacterium smegmatis* mc² 155

Elisângela F. Pimentel-Schmitt[a]   Knut Jahreis[b]   Mike P. Eddy[c]   Johannes Amon[a]
Andreas Burkovski[a]   Fritz Titgemeyer[a]

[a] Department of Microbiology, Friedrich Alexander University Erlangen-Nürnberg, Erlangen;
[b] Department of Biology and Chemistry, University of Osnabrück, Osnabrück, Germany;
[c] BioTechHome.com, Del Mar, Calif., USA

Key Words
Sugar transport · Membrane topology · Mycobacterium · Pathogenicity · Bacterial lifestyle · Carbon regulation

Abstract
We report here the molecular identification of a glucose permease from *Mycobacterium smegmatis*, a model organism for our understanding of the life patterns of the major pathogens *Mycobacterium tuberculosis* and *Mycobacterium leprae*. A computer-based search of the available genome of *M. smegmatis* mc² 155 with the sequences of well-characterized glucose transporters revealed the gene *msmeg4187* as a possible candidate. The deduced protein belongs to the major facilitator superfamily of proton symporters and facilitators and exhibits up to 53% of amino acid identity to other members of this family. Heterologous expression of *msmeg4187* in an *Escherichia coli* glucose-negative mutant led to the restoration of growth on glucose. The determination of the biochemical features characterize MSMEG4187 (GlcP) as a high affinity (K_m of 19 μM), glucose-specific permease. The results represent the first molecular characterization of a sugar permease in mycobacteria, and thus supply fundamental data for further in-depth analysis on the nutritional lifestyle of these bacteria.

Copyright © 2008 S. Karger AG, Basel

Introduction

Mycobacterium tuberculosis remains a serious threat to human health [Kaufmann, 2006]. About one-third of the world's population is thought to be infected with this pathogenic microorganism. In spite of significant investments in research, the mechanisms of its pathogenicity are still not fully understood. One aspect that is inextricably linked to pathogenicity is nutrition. This has been recently demonstrated by the identification of central metabolic genes, including carbohydrate transporter genes, which are essential for entering into and proliferating inside the macrophage [Joshi et al., 2006]. The current literature reveals limited information about the intermediary metabolism, utilization of organic compounds in the process of growth, and carbon catabolite control of mycobacteria. Efforts have been undertaken to understand virulence in mycobacteria by using the fast-growing *Mycobacterium smegmatis*. This model organism is infrequently reported as pathogenic and causes skin or soft tissue infections [Brinton et al., 1991; Lustgarten, 1884].

We reported previously the identification of a glucose kinase gene from *M. smegmatis* mc² 155, which could be a potential signalling factor for carbon regulation [Pimentel-Schmitt et al., 2006]. Since the transport of a carbon source represents, for many pathways, the decisive step for the flux through the pathway [Ruyter et al., 1991], we aimed to unravel the glucose uptake system(s). A putative glucose transporter that is encoded by the gene *msmeg4187* was recently detected by us in the course of a genome analysis of the carbohydrate permeases of *M. smegmatis* and *M. tuberculosis* [Titgemeyer et al., 2007].

In this communication, we demonstrate the function of the possible glucose transporter by heterologous complementation in an appropriate mutant of *Escherichia coli*. Furthermore, we provide biochemical details on the catalytic function of GlcP (MSMEG4187) and present a detailed topology model that includes the coordinates of conserved amino acid residues among related bacterial glucose permeases. GlcP^{Msm} is to the best of our knowledge the first molecular described sugar permease in the genus mycobacteria.

Results

msmeg4187 Encodes a Putative Glucose Permease

The protein database of the *M. smegmatis* genome was scanned with several known glucose permease protein sequences as implemented at The Institute for Genomic Research (TIGR) [Parche et al., 2006; van Wezel et al., 2005]. We found only one gene (*msmeg4187*) that encodes a predicted glucose permease of the major facilitator super family (MFS) [Pao et al., 1998]. The deduced protein shares 53% amino acid identity to the well-characterized glucose permease GlcP of *Streptomyces coelicolor* [Titgemeyer et al., 2007; van Wezel et al., 2005]. Analysis of the gene locus of *msmeg4187* (hereafter named *glcP*) indicated that the gene is transcribed as a monocistronic mRNA (fig. 1). *glcP* is flanked downstream by a putative dicistronic *hisEG* operon of genes involved in histidine biosynthesis, and upstream by an unknown open reading frame. While the *hisEG* operon is conserved in mycobacteria, it appears that the slow-growing mycobacteria, e.g. *M. tuberculosis, M. bovis*, and *M. avium*, do not have an obvious homologue of *glcP*.

The multiple alignment shown in figure 1b includes four homologous glucose permeases which have been experimentally described [Parche et al., 2006; van Wezel et al., 2005; Weisser et al., 1995; Zhang et al., 1989]. The range of protein identity is from 34 to 53%, showing 67 conserved residues in these sequences. The two typical signature sequences of the MFS are found in GlcP^{Msm} at positions 126–151 (VgGIGvGVasviapayiaETsppgiR) and 331 to 348 (AIALIDKIGRKpllligS). Hydrophobicity profiles showed that all five sequences have 12 predicted transmembrane helices that are at equivalent positions (fig. 1b). To establish the phylogenetic relationships of GlcP^{Msm}, we selected 19 protein sequences of the MFS which have been characterized by experimental investigation and those from diverse mycobacteria (fig. 1C). GlcP^{Msm} clusters with the *S. coelicolor* homolog and the one from the cyanobacterium *Synechocystis* [van Wezel et al., 2005; Zhang et al., 1989]. Interestingly, the xylose permease XylE of *E. coli* is more closely related to GlcP^{Msm} than expected, while the bifidobacterial GlcP is more distant but closely associated with another xylose permease, XylT of *Lactobacillus brevis* [Chaillou et al., 1998]. This suggests that glucose and xylose transporter genes may have evolved following two different branches from the common origins.

Heterologous Expression of GlcP in E. coli

The sequence of *msmeg4187* was amplified from chromosomal DNA and cloned into plasmid pSU2719 in orientation of the *lacZ* gene, yielding plasmid pFT283 (*glcP*$^+$). The plasmid was introduced into *E. coli* LJ140 (Δ*ptsHIcrr::kan*), which does not transport glucose to appreciable

Fig. 1. Genetic organization of the *glcP* gene in *M. smegmatis*, protein alignment, and phylogenetic tree. **a** Genetic map. The depicted region of *glcP* (black) in *M. smegmatis* includes the genes *hisG, hisE* (both histidine biosynthetic genes), and a gene encoding a protein of unknown function (*urf*, MSMEG4188). The genome of *M. smegmatis* (mc^2 155) was sequenced by TIGR (http://www.tigr.org/). The arrows indicate length and transcriptional orientation of annotated genes which are depicted by the respective *orf* number. Numbers in square brackets show the lengths of intergenic regions in bp. **b** The amino acid sequences of *glcP* in *M. smegmatis* were aligned with representative homologues. Residues identical in all proteins are shaded black and those that are conserved are shown in grey. Above the alignment, the residues present in all proteins are marked by asterisks. Abbreviations and protein designations are as follows. Msm, *M. smegmatis* (MSMEG4187); Sco, *S. coelicolor* (SCO5578); Syn, *Synechocystis* sp. PCC6803 (sll0771); Blo, *B. longum* (BL1631); Zmo, *Z. mobilis* (ZMO0366). **c** Phylogenetic tree of the *glcP* gene. An unrooted phylogenetic tree was computed with the CLUSTALW software making use of the implemented neighbor joining method with the function for evolutionary distance correction. Evolutionary distances are proportional to the branch length. 19 protein sequences were selected as indicated in the figure. The references for experimentally verified transporters are as follows: Chaillou et al., 1998; Krispin and Allmansberger, 1998; Martin et al., 1994; Parche et al., 2006; van Wezel et al., 2005; Weisser et al., 1995; Zhang et al., 1989.

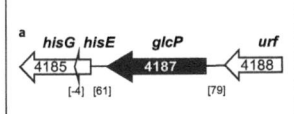
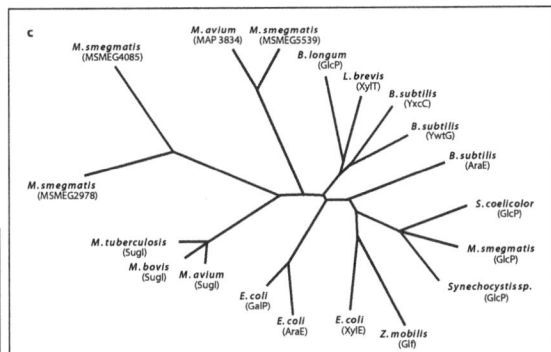
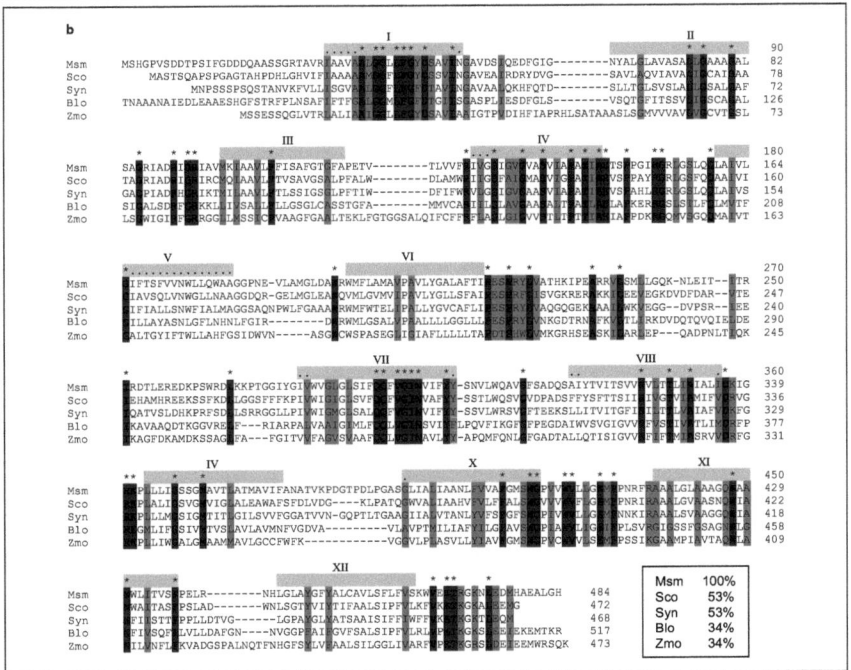

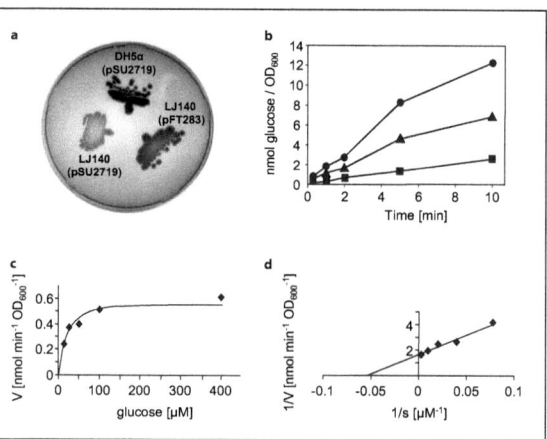

Fig. 2. Heterologous complementation and tranport kinetics. **a** Complementation of the glucose-negative phenotype in *E. coli* LJ140 (pFT283 *glcP*⁺) is shown on a MacConkey agar base plate supplemented with 50 mM glucose. Pigmented red (here grey) colonies indicate the fermentation of glucose, whereas white colonies reflect a deficiency in glucose fermentation. DH5α (pSU2719) was used as a positive control (dark gray) and LJ140 (pSU2719) as the congenic negative control. **b** Time-dependent uptake of 100 μM [^{14}C]glucose using cells of DH5α(pSU2719) (circles, positive control), LJ140 (pFT283) (triangles) and LJ140 (pSU2719) negative control (squares). Standard deviations of triplicate data points are shown by error bars but are barely visible due to minimal data variation. **c** [^{14}C]glucose uptake at different concentrations were fit to a non-linear regression and (**d**) graphed into a Lineweaver-Burk plot revealing a K_m of 19 μM and V_{max} of 0.63 nmol min^{-1} OD$_{600}$$^{-1}$.

amounts due to a chromosomal deletion of the *pts* operon. LJ140(pFT283), and several control strains were streaked onto MacConkey agar plates supplemented with 50 mM glucose (fig. 2a). The presence of *M. smegmatis glcP* led to the restoration of glucose fermentation as indicated by the formation of gray colonies. In comparison, the congenic strain LJ140(pSU2719) that carries the same cloning vector without the *glcP* gene grew as white colonies.

GlcP Is a High-Affinity Glucose Permease with Narrow Substrate Specificity

Glucose uptake experiments were performed to further characterize the biochemical features of the GlcP^{Msm} permease (fig. 2b). While LJ140(pSU2719) showed negligible uptake kinetics, transport of glucose was readily detectable when *msmeg4187* was expressed in LJ140(pFT283), reaching about 50% of the intrinsic *E. coli* wild-type glucose transport activity that was monitored in *E. coli* DH5α(pSU2719). A series of uptake experiments with various glucose concentrations was then performed to determine the kinetic parameters of GlcP (fig. 2c, d). We determined an affinity constant (K_m) of 19 μM and a velocity (V_{max}) of 0.63 nmol min^{-1} OD$_{600}$$^{-1}$. The substrate specificity was investigated by a competitive inhibition of glucose transport. Therefore, we added a hundredfold excess (10 mM) of nonradioactive sugar. While the presence of excess glucose and the derivative 2-deoxyglucose led to a reduction of more than 80% of the initial uptake rate, no or very small effects (<10%) were observed, when methyl α-D-glucopyranoside, xylose, arabinose, fructose, and trehalose were tested. In conclusion, the biochemical analysis of GlcP^{Msm} revealed a high-affinity permease with a narrow substrate specificity.

Discussion

The MFS family is one of the largest transporter families with members in bacteria and in eukaryotes like the GLUT facilitator transporters of mammals [Pao et al., 1998]. The transmembrane proteins of the GLUT family exhibit different substrate specificities, kinetic properties and tissue expression profiles. GLUT1 for example is expressed as a high-affinity permease in erythrocytes, the perineurium of peripheral nerves, and capillary endothelial cells of the blood-brain barrier [Gould et al., 1991; Mueckler et al., 1985; Salas-Burgos et al., 2004]. GLUT2 is a low-affinity glucose transporter, expressed in the liver, pancreas, intestine, and kidneys [Fukumoto et al., 1988; Gould et al., 1991]. GLUT3 and GLUT4 are high-affinity symporters present in brain, heart, muscle, and brown adipose tissue [Fukumoto et al., 1989; Kayano et

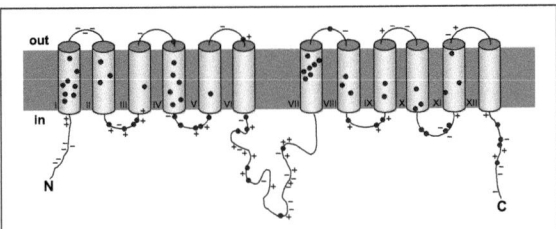

Fig. 3. Topology model of *M. smegmatis* GlcP. The drawing displays the predicted two-dimensional topology structure of GlcP, which is composed of 12 putative transmembrane segments. Conserved residues present in bacterial GlcPs (fig. 1b) are highlighted by black circles. Charged amino acids are marked by (+) and (–), showing a charge distribution of more residues on the inside of a transmembrane helix compared to the outside as defined by the positive-inside rule described by Heijne [1986].

al., 1988; Wood and Trayhurn, 2003]. Thus, this type of glucose permease has obviously been very successful in evolution.

In this study, we report by molecular means the first mycobacterial sugar permease and established that heterologous gene expression of a mycobacterial integral membrane protein in *E. coli* is feasible. The permease is the glucose-specific transport protein GlcP from *M. smegmatis* mc^2 155 encoded by open reading frame *msmeg4187*. GlcP^{Msm} exhibited an apparent K_m of 19 μM when expressed in *E. coli*. This correlates with the corresponding values of 41 μM of *S. coelicolor*'s GlcP [van Wezel et al., 2005] and 70 μM of *B. longum*'s GlcP [Parche et al., 2006]. GlcP^{Msm} shows the closest phylogenetic relationship to glucose permeases from streptomycetes and cyanobacteria [van Wezel et al., 2005; Zhang et al., 1989], which differ in their substrate specificity for fructose. GlcP^{Msm} was found specific for glucose and 2-deoxyglucose and not for fructose. Thus, it resembles in its substrate specificity profile the homolog GlcP^{Sco} from *S. coelicolor* [van Wezel et al., 2005]. In fact, our recent genome-based compilations of all carbohydrate transport systems of *S. coelicolor* and *M. smegmatis* shows that these two soil-dwelling actinomycetes have many homologous sugar uptake systems in common [Bertram et al., 2004; Titgemeyer et al., 2007].

Our bioinformatic analysis on the membrane topology of GlcP^{Msm} led to the model presented in figure 3. GlcP^{Msm} has a predicted composition of 12 transmembrane helices typical for MFS porters [Pao et al., 1998]. The model was obtained by aligning the predictions of possible transmembrane segments of the five GlcP proteins depicted in figure 1b assuming that they all share the same overall topology. This showed that 12 transmembrane segments are probably present in all of them, while sequence variations in terms of deletions and insertions occurred only in regions which were predicted as flexible loops. Figure 3 displays that helices I, IV, and VII exhibit (with eight, six, and seven conserved residues) the highest conservation suggesting that these may be most relevant for substrate recognition.

Interestingly, no GlcP homolog is present in slow-growing mycobacteria, notably *M. tuberculosis* and *M. leprae* [Titgemeyer et al., 2007]. In these bacteria, glucose may enter either by simple diffusion, which could be efficient enough to meet the nutritional demands of these extremely slow-growing bacteria (generation time of weeks), or through another, yet unknown permease [Titgemeyer et al., 2007]. *M. smegmatis* and also the slow-growing mycobacterial species possess a glucose kinase that converts the incoming glucose to glucose-6-phosphate [Pimentel-Schmitt et al., 2006; Titgemeyer et al., 2007]. Apart from its catalytic function, this enzyme confers global carbon regulation in closely related streptomycetes [Angell et al., 1994]. We could recently show in *S. coelicolor* that glucose kinase binds to glucose permease GlcP [van Wezel et al., 2007]. This protein-protein interaction is thus thought to influence the regulatory function of glucose kinase. For this reason, it would be worthwhile to study, whether GlcP and GlkA of *M. smegmatis* also interact with each other and whether this has similar consequences regarding nutritional control. This in turn would help to understand how nutrition and virulence may be linked in mycobacteria.

Experimental Procedures

Bacterial Strains and Growth Conditions

Cells of *E. coli* were grown at 37°C under vigorous shaking in Luria-Bertani broth (LB) and supplemented with the appropriate antibiotics when required. *E. coli* DH5α served as the host strain for gene cloning purposes [Sambrook et al., 1989]. LJ140 (Δ*ptsHI::*

kan) is a derivative of *E. coli* LJ110 [Zeppenfeld et al., 2000], which a deletion of the *pts* operon was introduced by common P1(kc) transduction through selection of a gene for kanamycin resistance which substitutes the *pts* locus.

Heterologous Complementation

Chromosomal DNA was prepared from 100-ml cultures grown to saturation [Belisle and Sonnenberg, 1998]. 400 ng was used as template for a polymerase chain reaction with oligonucleotides FS12 ggctgtctagatcattgggccgcaggaagaaga and FPS21 tgcgccgatgggctctagatgccgtcagt (engineered *Hin*dIII and *Xba*I restriction sites are underlined). Primer FS12 contained stop codons in all three reading frames and an artificial ribosome binding site for *glcP* to ensure efficient translation. The amplified DNA fragment was digested and ligated in pSU2719 in the same orientation as the *lacZ* gene [Chandler, 1991]. The ligation mixture was transformed into DH5α. Recombinant plasmids were isolated and confirmed by DNA sequencing. The resulting *glcP*^{Msm} expression vector was designated pFT283. LJ140 was transformed with pFT283 or pSU2719 (control), and transformant colonies were streaked out on MacConkey agar plates supplemented with 50 mM glucose in order to examine heterologous complementation of the glucose transport-negative phenotype.

Transport Assays

Cells of *E. coli* LJ140 bearing either plasmid pSU2719 or pFT283 *glcP*$^+$ and *E. coli* DH5α(pSU2719) were grown at 37°C in 20 ml LB supplemented with chloramphenicol to an OD$_{600}$ of 1.0. Cells were pre-equilibrated to 37°C before uptake was initiated by the addition of [^{14}C]glucose at a final concentration of 100 μM (100,000 cpm/ml) [Nothaft et al., 2003]. Samples of 0.5 ml were taken at different time points, rapidly filtered through nitrocellulose filters, and washed with 5 ml chilled 0.1 M LiCl. Radioactivity was determined by scintillation oscillography with 3 ml scintillation solution per filter. Uptake rates are expressed in nanomoles min^{-1} OD$_{600}$$^{-1}$. For determination of the kinetic parameters, glucose uptake experiments were conducted with various glucose concentrations (12.5–400 μM). To determine V_{max} and K_m, the data were fit to the Michaelis-Menten equation by a nonlinear regression program and graphed on a Lineweaver-Burk plot. The substrate specificity was studied by glucose uptake competition assay using a 100-fold excess of nonradioactive sugar (10 mM) as described previously [van Wezel et al., 2005]. Transport activity without additional sugar was set to 100% (control).

Computational Analyses

The genome data of *M. smegmatis* mc^2 155 were collected from the primary annotation database (http://cmr.tigr.org/tigr-scripts/CMR/shared/Genomes.cgi) available at The Institute for Genomic Research (TIGR). Protein databank searches were carried out at the BLAST server of the National Center for Biotechnology Information at the National Institutes of Health, Bethesda, Md., USA (http://www.ncbi.nlm.nih.gov), the BLAST server of the Transport Classification Databank, TCDB (www.tcdb.org), and at TIGR. Multiple alignment analysis and graphical optimization were achieved with CLUSTALW (www.ebi.ac.uk/clustalw) and BOXSHADE (www.ch.embnet.org/software/BOX_form.html) software. Hydrophobicity profiles and topology predictions were calculated with the TopPred II software [Claros and von Heijne, 1994]. The phylogenetic tree was calculated with the neighbor joining method as implemented in CLUSTALW. The tree was graphically visualized by importing phylip (*.ph) files into the TREEVIEW (http://taxonomy.zoology.gla.ac.uk/rod/treeview.html) software that was kindly provided by Rod Page.

Acknowledgements

This study was supported through grants of SFB473 to F.T., GK805 to E.F.P.-S., and SFB431 to K.J. of the Deutsche Forschungsgemeinschaft.

References

Angell S, Lewis CG, Buttner MJ, Bibb MJ: Glucose repression in *Streptomyces coelicolor* A3(2): a likely regulatory role for glucose kinase. Mol Gen Genet 1994;244:135–143.

Belisle JT, Sonnenberg MG: Isolation of genomic DNA from mycobacteria. Methods Mol Biol 1998;101:31–44.

Bertram R, Schlicht M, Mahr K, Nothaft H, Saier MH Jr, Titgemeyer F: In silico and transcriptional analysis of carbohydrate uptake systems of *Streptomyces coelicolor* A3(2). J Bacteriol 2004;186:1362–1373.

Brinton LA, Li JY, Rong SD, Huang S, Xiao BS, Shi BG, Zhu ZJ, Schiffman MH, Dawsey S: Risk factors for penile cancer: results from a case-control study in China. Int J Cancer 1991;47:504–509.

Chaillou S, Bor YC, Batt CA, Postma PW, Pouwels PH: Molecular cloning and functional expression in *Lactobacillus plantarum* 80 of *xylT*, encoding the D-xylose-H+ symporter of *Lactobacillus brevis*. Appl Environ Microbiol 1998;64:4720–4728.

Chandler MS: New shuttle vectors for *Haemophilus influenzae* and *Escherichia coli*: P15A-derived plasmids replicate in *H. influenzae* Rd. Plasmid 1991;25:221–224.

Claros MG, von Heijne G: TopPred II: an improved software for membrane protein structure predictions. Comput Appl Biosci 1994;10:685–686.

Fukumoto H, Seino S, Imura H, Seino Y, Eddy RL, Fukushima Y, Byers MG, Shows TB, Bell GI: Sequence, tissue distribution, and chromosomal localization of mRNA encoding a human glucose transporter-like protein. Proc Natl Acad Sci USA 1988;85:5434–5438.

Fukumoto H, Kayano T, Buse JB, Edwards Y, Pilch PF, Bell GI, Seino S: Cloning and characterization of the major insulin-responsive glucose transporter expressed in human skeletal muscle and other insulin-responsive tissues. J Biol Chem 1989;264:7776–7779.

Gould GW, Thomas HM, Jess TJ, Bell GI: Expression of human glucose transporters in *Xenopus* oocytes: kinetic characterization and substrate specificities of the erythrocyte, liver, and brain isoforms. Biochemistry 1991;30:5139–5145.

Heijne G: The distribution of positively charged residues in bacterial inner membrane proteins correlates with the trans-membrane topology. EMBO J 1986;5:3021–3027.

Joshi SM, Pandey AK, Capite N, Fortune SM, Rubin EJ, Sassetti CM: Characterization of mycobacterial virulence genes through genetic interaction mapping. Proc Natl Acad Sci USA 2006;103:11760–11765.

Kaufmann SH: Envisioning future strategies for vaccination against tuberculosis. Nat Rev Immunol 2006.

Kayano T, Fukumoto H, Eddy RL, Fan YS, Byers MG, Shows TB, Bell GI: Evidence for a family of human glucose transporter-like proteins: sequence and gene localization of a protein expressed in fetal skeletal muscle and other tissues. J Biol Chem 1988;263:15245–15248.

Krispin O, Allmansberger R: The *Bacillus subtilis* AraE protein displays a broad substrate specificity for several different sugars. J Bacteriol 1998;180:3250–3252.

Lustgarten S: Über spezifische Bacillen in syphilitischen Krankheitsprodukten. Wienerische Medizinische Wochenschrift 1884:1.

Martin GE, Seamon KB, Brown FM, Shanahan MF, Roberts PE, Henderson PJ: Forskolin specifically inhibits the bacterial galactose-H$^+$ transport protein, GalP. J Biol Chem 1994;269:24870–24877.

Mueckler M, Caruso C, Baldwin SA, Panico M, Blench I, Morris HR, Allard WJ, Lienhard GE, Lodish HF: Sequence and structure of a human glucose transporter. Science 1985; 229:941–945.

Nothaft H, Parche S, Kamionka A, Titgemeyer F: In vivo analysis of HPr reveals a fructose-specific phosphotransferase system that confers high-affinity uptake in *Streptomyces coelicolor*. J Bacteriol 2003;185:929–937.

Pao SS, Paulsen IT, Saier MH Jr: Major facilitator superfamily. Microbiol Mol Biol Rev 1998; 62:1–34.

Parche S, Beleut M, Rezzonico E, Jacobs D, Arigoni F, Titgemeyer F, Jankovic I: Lactose-over-glucose preference in *Bifidobacterium longum* NCC2705:*glcP*, encoding a glucose transporter, is subject to lactose repression. J Bacteriol 2006;188:1260–1265.

Pimentel-Schmitt EF, Thomae AW, Amon J, Klieber AM, Roth H-M, Muller YA, Jahreis K, Burkovski A, Titgemeyer F: A glucose kinase from *Mycobacterium smegmatis*. J Mol Microbiol Biotechnol 2007;12:75–81.

Ruyter GJ, Postma PW, van Dam K: Control of glucose metabolism by enzyme II^{Glc} of the phosphoenolpyruvate-dependent phosphotransferase system in *Escherichia coli*. J Bacteriol 1991;173:6184–6191.

Salas-Burgos A, Iserovich P, Zuniga F, Vera JC, Fischbarg J: Predicting the three-dimensional structure of the human facilitative glucose transporter glut1 by a novel evolutionary homology strategy: insights on the molecular mechanism of substrate migration, and binding sites for glucose and inhibitory molecules. Biophys J 2004;87:2990–2999.

Sambrook J, Fritsch EF, Maniatis T: Molecular Cloning: A Laboratory Manual, ed 2. Cold Spring Harbor, Cold Spring Harbor Laboratory Press, 1989.

Titgemeyer F, Amon J, Parche S, Mahfoud M, Bail J, Schlicht M, Rehm N, Hillmann D, Stephan J, Walter B, Burkovski A, Niederweis M: A genomic view of sugar transport in *Mycobacterium smegmatis* and *Mycobacterium tuberculosis*. J Bacteriol 2007;189: 5903–5915.

van Wezel GP, Mahr K, König M, Traag BA, Pimentel-Schmitt EF, Willimek A, Titgemeyer F: GlcP constitutes the major glucose uptake system of *Streptomyces coelicolor* A3(2). Mol Microbiol 2005;55:624–636.

van Wezel GP, König M, Mahr K, Nothaft H, Thomae AW, Bibb MJ, Titgemeyer F: A new piece of an old jigsaw: glucose kinase is activated posttranslationally in a glucose transport-dependent manner in *Streptomyces coelicolor* A3 (2). J Mol Microbiol Biotechnol 2007;12:65–72.

Weisser P, Krämer R, Sahm H, Sprenger GA: Functional expression of the glucose transporter of *Zymomonas mobilis* leads to restoration of glucose and fructose uptake in *Escherichia coli* mutants and provides evidence for its facilitator action. J Bacteriol 1995;177:3351–3354.

Wood IS, Trayhurn P: Glucose transporters (GLUT and SGLT): expanded families of sugar transport proteins. Br J Nutr 2003;89: 3–9.

Zeppenfeld T, Larisch C, Lengeler JW, Jahreis K: Glucose transporter mutants of *Escherichia coli* K-12 with changes in substrate recognition of IICB^{Glc} and induction behavior of the *ptsG* gene. J Bacteriol 2000;182:4443–4452.

Zhang CC, Durand MC, Jeanjean R, Joset F: Molecular and genetical analysis of the fructose-glucose transport system in the cyanobacterium *Synechocystis* PCC6803. Mol Microbiol 1989;3:1221–1229.

A Glucose Kinase from *Mycobacterium smegmatis*

Elisângela F. Pimentel-Schmitt[a]   Andreas W. Thomae[b]   Johannes Amon[a]
Michael A. Klieber[c]   Heide-Marie Roth[c]   Yves A. Muller[c]   Knut Jahreis[d]
Andreas Burkovski[a]   Fritz Titgemeyer[a]

[a]Department of Microbiology, Friedrich Alexander University Erlangen-Nürnberg, Erlangen; [b]Department of Gene Vectors, GSF-National Research Center for Environment and Health, Munich; [c]Lehrstuhl für Biotechnik, Universität Erlangen-Nürnberg IZMP, Erlangen; [d]Department of Biology and Chemistry, University of Osnabrück, Osnabrück, Germany

Key Words
Carbon regulation · Glucose metabolism · Catabolite repression · Hexokinase · Glucokinase

Abstract
Carbon metabolism and regulation is poorly understood in mycobacteria, a genus that includes some major pathogenic species like *Mycobacterium tuberculosis* and *Mycobacterium leprae*. Here, we report the identification of a glucose kinase from *Mycobacterium smegmatis*. This enzyme serves in glucose metabolism and global carbon catabolite repression in the related actinomycete *Streptomyces coelicolor*. The gene, *msmeg1356* (*glkA*), was found by means of in silico screening. It was shown that it occurs in the same genetic context in all so far sequenced mycobacterial species, where it is located in a putative tricistronic operon together with a glycosyl hydrolase and a putative malonyl-CoA transacylase. Heterologous expression of *glkA* in an *Escherichia coli* glucose kinase mutant led to the restoration of glucose growth, which provided in vivo evidence for glucose kinase function. GlkA^{Msm} was subsequently overproduced in order to study its enzymatic features. We found that it can form a dimer and that it efficiently phosphorylates glucose at the expense of ATP. The affinity constant for glucose was with 9 mM about eight times higher and the velocity was about tenfold slower when compared to the parallel measured glucose kinase of *S. coelicolor*. Both enzymes showed similar substrate specificity, which consists in an ATP-dependent phosphorylation of glucose and no, or very inefficient, phosphorylation of the glucose analogues 2-deoxyglucose and methyl α-glucoside. Hence, our data provide a basis for studying the role of mycobacterial glucose kinase in vivo to unravel possible catalytic and regulatory functions.   Copyright © 2007 S. Karger AG, Basel

E.F.P.-S. and A.W.T contributed equally to this study.

Introduction

The re-emergences of diseases caused by tuberculosis and non-tuberculosis mycobacteria have prompted primary medical interest in these organisms. The mycobacterial genus contains besides *Mycobacterium tuberculosis*, the human pathogens *Mycobacterium leprae*, and *Mycobacterium ulcerans* [Reed et al., 2006]. These bacteria cause tuberculosis with approximately 2 million fatal casualties each year [Sharma et al., 2006], leprosy, and severe skin-destructive diseases. Furthermore, *Mycobacterium avium* and *Mycobacterium intracellulare* are of actual medical relevance since they are to be found in human immunodeficiency virus-infected individuals

[Reed et al., 2006]. Consequently, current research efforts are directed towards an understanding of the underlying virulence mechanisms. Little is known about the basic biochemical information in mycobacteria, although this knowledge is often essential for the comprehension of the pathogenic life cycle, and therefore, for the development of new therapeutics [Waagmeester et al., 2005].

One of the critical factors for the success of *M. tuberculosis* as a pathogen appears to be its ability for differential expression of genes required for its survival under various conditions [Agarwal and Tyagi, 2006]. Considering that glucose is a preferred source of carbon and energy for many bacteria, the transport of this sugar into the cell and its subsequent phosphorylation is related to global gene regulation, and may also influence pathogenicity [Deol et al., 2005; Mathur et al., 2005]. *M. tuberculosis* virulence is correlated with a shift from a strict aerobic respiratory mode to anaerobic metabolism. Although expression of glycolytic enzymes is necessary for steady-state metabolism, the coordinated increased expression of genes encoding glycolytic enzymes is particularly important for adaptation to anaerobiosis [Smith and Neidhardt, 1983]. Growth studies have indicated that up to 70% of glucose is metabolized through the Embden-Meyerhof-Parnas pathway [Jayanthi Bai et al., 1975; Ramakrishnan et al., 1962]. Therefore, glycolysis is central to the organism's survival and consequently a potential drug target [Mathur et al., 2005].

Glucose kinase is the enzyme that funnels incoming glucose into glycolysis by converting glucose to glucose-6-phosphate. Glucose is also a key nutrient for cell wall biosynthesis in mycobacteria, as glucose-6-phosphate is required for the galactan residue of the arabinogalactan cell wall layer [Huang et al., 2006; Trejo et al., 1971]. Furthermore, glucose kinase plays a central role in global carbon regulation in the closely related actinomycete *Streptomyces coelicolor*. This reinforces the importance of glucose kinase in gene expression [Angell et al., 1994; Kwakman and Postma, 1994; Ramos et al., 2004; van Wezel et al., 2006].

For this reason, it was our goal to identify and characterize a glucose kinase from mycobacteria. We therefore screened the sequences of several mycobacterial genomes and selected a putative glucose kinase gene from the fast-growing non-pathogenic *Mycobacterium smegmatis*. *M. smegmatis*, which is infrequently reported as infectious

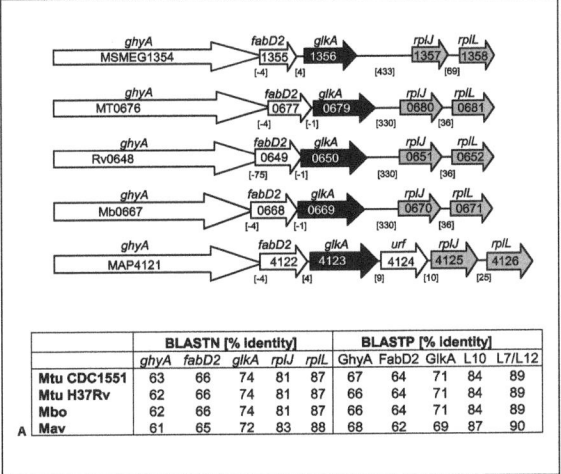

Fig. 1. Genetic organization, multiple alignment and phylogenetic tree. **A** Genetic organization of the mycobacterial *glkA* gene locus. The highly conserved depicted *glkA* region (black) includes the genes *ghyA* (encoding a glycosyl hydrolase), *fabD2* (malonyl-CoA transacylase), and the genes for the ribosomal proteins L10 and L7/L12 (gray). Note that *M. avium paratuberculosis* additionally carries a gene encoding a protein of unknown function (*urf*, MAP4124). The genomes of *M. smegmatis* mc² 155, *M. tuberculosis* H37Rv, *M. tuberculosis* CDC1551 *M. bovis* subsp. *bovis* AF2122/97, and *M. avium paratuberculosis* were sequenced and annotated by TIGR (http://www.tigr.org/). The arrows indicate length and transcriptional orientation of annotated genes and open reading frames. Numbers in brackets show the lengths of intergenic regions in bp. The gene names are assigned according to the annotations given by TIGR and by us. The BLAST (http://tigrblast.tigr.org/cmr-blast/) results for the genes given in the additional table are [% identity] for nucleotide and protein BLAST, respectively.

in humans and in animals [Bercovier et al., 2001], is commonly used as a model organism for this genus [Wang et al., 2005]. We show that the predicted glucose kinase gene *msmeg1356* is also found in a conserved genetic environment in other mycobacteria, including the *M. tuberculosis* reference strain H37Rv. The gene was cloned and overproduced in *Escherichia coli*. The purified protein exhibited all biochemical features of a glucose kinase. In addition, we demonstrate its physiological function by heterologous complementation of a glucose kinase *E. coli* mutant.

Results

Prediction of a Possible Glucose Kinase Gene

The genomes of *M. smegmatis* mc² 155 and *M. tuberculosis* H37RV were screened for the presence of putative glucose kinase genes. A protein BLASTP search with the Glk protein sequence of *S. coelicolor* revealed the presence of seven homologues in the genome of *M. smegmatis*. They are members of the ROK family, which comprises of bacterial transcription factors and sugar kinases [Titgemeyer et al., 1994b]. Three of the seven genes are

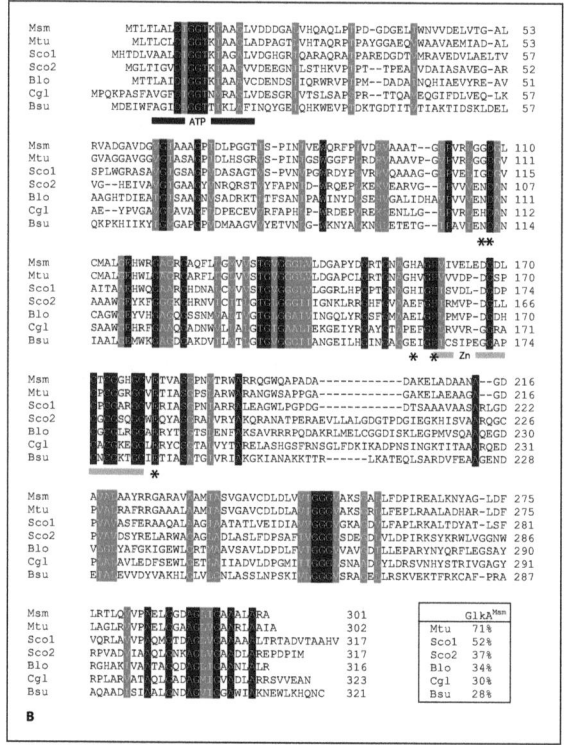

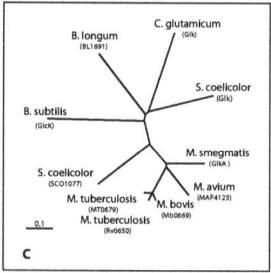

Fig. 1. B The amino acid sequence of GlkA^{Msm} was aligned with representative homologues. Residues conserved in all proteins are shaded black (identical) and gray (similar). Below the alignment functionally important regions and residues are indicated by black bars (ATP-binding site), gray bars (zinc-binding site), and asterisks (amino acids involved in glucose recognition). Abbreviations and accession numbers are as follows: Msm, *M. smegmatis* mc² 155 (MSMEG1356); Mtu, *M. tuberculosis* CDC1551 (MT0679); Sco1, *S. coelicolor* (SCO1077); Sco2, *S. coelicolor* (SCO212, *glk*); Blo, *B. longum* (BL1691); Cgl, *C. glutamicum* (CG2399); Bsu, *B. subtilis* (BSU2483). **C** An unrooted phylogenetic tree is depicted that was computed with the CLUSTALW software using the implemented neighbor joining method. Evolutionary distances are proportional to the branch length. Names of proteins and species are indicated.

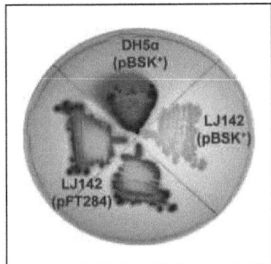

Fig. 2. Functional complementation of GlkA^{Msm}. Complementation of the glucose-negative phenotype in *E. coli* LJ142(pFT284 *glkA*$^+$) is shown on a MacConkey agar base plate supplemented with 50 mM glucose. Pigmented red colonies indicate the fermentation of glucose, whereas white colonies reflect a deficiency in glucose fermentation. DH5α/pBSK$^+$ was used as a positive control and LJ142(pBSK$^+$) as the congenic negative control.

sugar kinases. Of these, the gene product MSMEG1356 exhibited with 37% identical and 58% similar amino acid residues the closest homologue. When we analyzed the genome of *M. tuberculosis*, we found three ROK family members, of which one gene, Rv0650, was close to *glk* of *S. coelicolor* and MSMEG1356. Hence, MSMEG1356 (hereafter termed *glkA*) was chosen as the prime candidate for a possible glucose kinase.

A comparison of the *glkA* region with other mycobacterial genomes revealed a very similar genetic context (fig. 1A), in which *glkA* is the third gene of a putative tricistronic operon. The first one is a possible glycosyl hydrolase, suggesting that the enzyme together with GlkA participates in the same pathway, which could be the metabolism of glucose-containing saccharides. The second gene appears to encode a putative malonyl-CoA transacylase found as a unique conserved mycobacterial gene [Huang et al., 2006]. Two genes that encode the ribosomal proteins L10 and L7/L12 are located immediately downstream of *glkA*. Such a conserved gene order is a good hint for equivalent functions in closely related species.

A multiple alignment of homologous glucose kinases from diverse bacteria reveals that GlkA^{Msm} shows 71% protein identity to the one from the *M. tuberculosis* H37Rv reference strain and 28–52% to the others that were chosen from related actinomycetes and from *Bacillus subtilis* (fig. 1B). GlkA^{Msm} contains the two ROK family signature sequences [LIVM]xxG[LIVMFCT]Gx[GA]xGx$_{3-5}$[GATP]xxG[RKH] found at position 136–151 and CxCG xxGx[WILV]Ex[YFVIN]x[STAG] at position 171–185 [Hansen et al., 2002; Titgemeyer et al., 1994a]. The latter motif is a cysteine-rich motif that is involved in zinc-binding [Schiefner et al., 2005]. An ATP-binding motif (residues 5–19) of the ß-strand-loop-ß-strand motif could be assigned according to the crystal structure data of the *E. coli* glucokinase [Lunin et al., 2004]. This structure also provided information on the putative amino acid residues involved in glucose binding. Three out of five (D108, H160, E180) are present in GlkA^{Msm}. Furthermore, the alignment shows that the two missing residues ([GN]107, [HE]157) occurred in several of the other aligned glucose kinases.

The unrooted tree shown in figure 1C visualizes the phylogenetic relationships. GlkA^{Msm} is most closely related to the mycobacterial homologues, then to SCO1077 (GlkII) of *S. coelicolor* and then equally distant to other glucose kinases.

Heterologous Complementation

To determine the function of GlkA from *M. smegmatis*, the gene was cloned into pBluescript vector (pBSK$^+$) by fusion to the strong *lacZ* promoter. The plasmid pFT284 (*glk*$^+$) was transformed into the glucose kinase mutant *E. coli* LJ142 (Δ*ptsHIcrr::kan* Δ*glk::cat*), which exhibits a glucose-negative phenotype due to an additional deletion of the *pts* operon. The strain showed restoration of glucose fermentation when plated on MacConkey agar indicator plates supplemented with 50 mM glucose (fig. 2). While the control LJ142(pBSK) formed white colonies, the presence of pFT284 resulted in red colonies indicating sugar fermentation. The complementation was confirmed by measuring glucose kinase activity. The activity values obtained from crude cell extracts of LJ142(pFT284) were 544 ± 2 nmol glucose-phosphate mg Pr^{-1} min^{-1}, whereas extracts of LJ142(pBSK$^+$) exhibited the intrinsic background activity of the assay (<10 U). Hence, the results of the heterologous complementation established that *msmeg1356* encodes a glucose kinase.

Biochemical Features

The glucose kinases of *M. smegmatis* mc^2 155 and of *S. coelicolor* A3(2) were overproduced and purified as histidine-tagged proteins in *E. coli* FT1(pLysS), harboring plasmids pFT255 (*glkA*^{Msm}) and pFT61 (*glk*^{Sco}) (fig. 3A) [Mahr et al., 2000]. The yield was 1–2 mg/l culture in both cases. Histidine-tagged GlkA of *M. smegmatis* migrated on polyacrylamide gels corresponding to a molecular

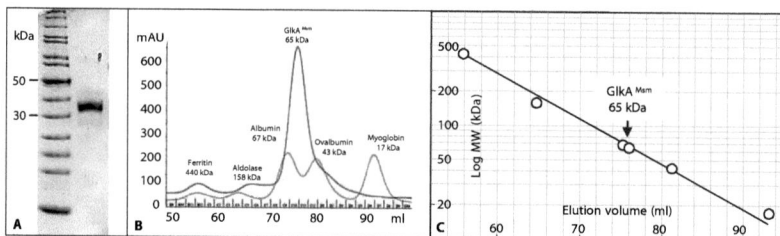

Fig. 3. Purification and oligomeric state of GlkA^{Msm}. **A** A Coomassie Brilliant Blue stained SDS-polyacrylamide gel shows homogenous purified His$_6$-GlkA^{Msm}. Molecular weights are indicated in kDa. **B** Size exclusion chromatography of the GlkA of *M. smegmatis*. Elution profile of the standard proteins ferritin (440 kDa), aldolase (158 kDa), albumin (67 kDa), ovalbumin (43 kDa), and myoglobin (17 kDa) were overlaid with the His$_6$-GlkA separation profile. **C** A calibration curve generated with elution volumes and molecular weights of standard proteins is shown. According to the elution volume of the His$_6$-GlkA^{Msm}, its native molecular weight was determined as a dimer of 65 kDa.

weight of about 33 kDa, which complies with a predicted molecular weight of 32,337 Da. Subsequently, analytical gel filtration was used to determine the native state of GlkA^{Msm} (fig. 3B, C). The derived size was 65 kDa, indicating that the protein is dimeric.

Glucose kinase activity assays were conducted to characterize the enzyme properties of GlkA^{Msm}. We found that the enzyme phosphorylates glucose in an ATP-dependent manner with a narrow specificity for the substrate glucose (table 1). This was demonstrated by substitution of glucose with the glucose analogues 2-deoxyglucose and methyl α-glucoside, which were not recognized by GlkA^{Msm}. Under the applied assay conditions, these features correspond to *S. coelicolor* his$_6$-Glk, which however exhibited low catalytic activity in the presence of methyl α-glucoside. The kinetic parameters for GlkA^{Msm} were inferred by monitoring the specific activity at various glucose concentrations. The maximum velocity of GlkA^{Msm} was 0.07 nmol glucose-phosphate mg Pr^{-1}min^{-1}, which was about tenfold slower compared to Glk^{Sco}. The affinity constants were 9 mM for GlkA^{Msm} and 1.2 mM for Glk^{Sco}. Thus, both enzymes showed similar properties.

Discussion

In this communication, we describe the first molecular characterization of a mycobacterial glucose kinase, a key enzyme for metabolism of glucose and glucose-con-

Table 1. Comparison of GlkA^{Msm} and GlkA^{Sco}

Substrate	GlkA^{Msm}	Glk^{Sco}
Glc	164 ± 0.9	1,351 ± 0.4
2-DOG	<4	<4
α-MG	<4	7 ± 0.2

The table shows the specific glucose kinase activity expressed in nmol glucose-phosphate min^{-1} mg protein^{-1}. 2-DOG = 2-Deoxyglucose; α-MG = methyl α-glucoside.

taining sugars, and a prime candidate for a signaling factor in global carbon regulation.

The gene *msmeg1356* of *M. smegmatis* was unraveled as a glucose kinase according to our in vivo and in vitro characterization. Its biochemical features are similar to those reported for the *S. coelicolor* glucose kinase, which acts in a second functional capacity as a signaling factor in global carbon regulation [see van Wezel et al., this issue, and Angell et al., 1994; Imriskova et al., 2005]. However, while GlkA^{Msm} is dimeric, it has been reported that Glk^{Sco} occurs in a tetrameric form [Imriskova et al., 2005]. Furthermore, the gene *sco1077* in the *S. coelicolor* genome appears with a 52% protein identity to be much closer related to GlkA^{Msm}. In contrast, the well-studied catalytic and regulatory glucose kinase encoded by *glk* shares only 37% identical amino acid residues with GlkA^{Msm}. SCO1077, however, seems not to be involved in carbon regulation [Angell et al., 1994].

The occurrence of a glucose kinase gene in all so far sequenced mycobacterial genomes indicates an important function. Our multiple alignment analysis revealed 39 amino acid residues, which are conserved in all six selected species. Among these residues are several that have been identified to participate in ATP, glucose, and zinc binding in distantly related species [Hansen et al., 2002; Lunin et al., 2004; Mesak et al., 2004; Schiefner et al., 2005; Titgemeyer et al., 1994a]. Furthermore, site-specific mutational analysis of *S. coelicolor* Glk concerning the ATP-binding site and the cysteine-rich motif resulted in every case in a complete loss of glucose kinase activity according to the model [pers. unpubl. data].

Future research should be focused on the question: which protein or metabolic pathways in *M. smegmatis* are affected by *glkA*? In particular, it will be important to examine whether *glkA* plays a direct metabolic role in glucose utilization and/or carbon regulation. So far, GlkA^{Msm} may not be required to maintain a functional virulence cycle. This at least can be inferred from a global microarray-based analysis of *M. tuberculosis* [Sassetti et al., 2003; Sassetti and Rubin, 2003]. Assuming that specific inhibitors of protein kinases have been successfully developed for therapeutic usage against a variety of diseases [Shawver et al., 2002], future investigation for drug development could be directed to the selection of GlkA of mycobacterium, as a means of interfering with growth and possible regulatory processes.

Experimental Procedures

Bacterial Strains and Growth Conditions

Cells of *E. coli* were grown at 37°C under vigorous shaking in Luria-Bertani broth (LB) and supplemented with the appropriate antibiotics when required. *E. coli* DH5α served as the host strain for gene cloning purposes [Sambrook et al., 1989]. LJ142 is a derivative of the *E. coli* K-12 wild-type LJ110 [Zeppenfeld et al., 2000], in which subsequently *glk::cat* and (*ptsHIcrr::kan*) mutations were introduced by common P1(kc) transductions. FT1 F(*ptsHIcrr::kan*) (pLysS) was used as the expression host for histidine-tagged glucose kinase genes [Parche et al., 1999].

Heterologous Complementation

To determine the function of GlkA from *M. smegmatis*, the *glkA* gene was cloned and amplified by complementation of *E. coli* LJ142, a glucose kinase-deficient mutant. For this purpose, *msmeg1356* (*glkA*) was amplified from chromosomal DNA of *M. smegmatis* mc^2 155 (kindly provided by Claudia Mailänder). The polymerase chain reaction was conducted with oligonucleotides FPS22 GCTCTAGAAATGACGCTCACCCTGGCCCT and FPS23 CGGAATTCGCCAGGGCGACGCGCAAACT engineered to introduce *Xba*I and *Eco*RI restriction sites (underlined). The amplified DNA fragment was digested and ligated in pBluescriptSK$^+$ behind the *lacZ* promoter. The ligation mixture was transformed into DH5α. Recombinant plasmids were isolated and confirmed by DNA sequencing. The *glkA*^{Msm} expression vector was designated pFT284. LJ142 was transformed with pFT284 or pBSK$^+$ (control), and transformant colonies were streaked out on MacConkey agar plates supplemented with 50 mM glucose to examine heterologous complementation of the glucose-negative phenotype. To confirm the complementation assay, glucose kinase activities were measured from crude cell extracts of LJ142(pBSK+) and LJ142(pFT284) as described [Skarlatos and Dahl, 1998].

Overproduction and Purification of Glucose Kinases

E. coli FT1(pLysS, pFT255 *glkA*$^{Msm+}$) and *E. coli* Rosetta(pFT61 *glk*$^{Sco+}$) [Mahr et al., 2000] were used for overproduction of histidine-tagged GlkA^{Msm} and histidine-tagged Glk^{Sco}. Overproduction and purification were performed as described [Mahr et al., 2000]. Cells were routinely grown in LB at 37°C to an OD$_{600}$ of 0.6 and gene expression was induced by addition of IPTG to a final concentration of 1 mM. Incubation was continued for 16 h for Glk of *S. coelicolor* and 4 h for the GlkA of *M. smegmatis*.

Size Exclusion Chromatography

A volume of 0.5 ml containing his$_6$-tagged GlkA^{Msm} (1 mg) or calibration proteins [myoglobin (1.5 mg/ml), ovalbumin (4 mg/ml), albumin (5 mg/ml), aldolase (0.2 mg/ml), ferritin (0.24 mg/ml)] was loaded onto an Äkta purifier Superdex 200 prep grade. All steps were carried out at 4°C. Fractions were collected and an absorption profile was generated. The size of native His$_6$-GlkA^{Msm} was calculated by linear regression analysis.

Glucose Kinase Assay

Glucose kinase activity was measured in a spectrophotometrically coupled assay by monitoring the reduction of NADP in a glucose-6-phosphate dehydrogenase coupled reaction. The assay mixture was composed of 1 μg of purified protein in a buffer containing 50 mM Tris/HCl, various glucose concentrations, 25 mM MgCl$_2$, 0.5 mM NADP, 1 mM ATP, and 0.7 U ml^{-1} glucose-6-phosphate dehydrogenase. The assay was conducted as previously described [Skarlatos and Dahl, 1998]. To determine V$_{max}$ and K$_m$ for each kinase, the data were fit to the Michaelis-Menten equation by a non-linear regression program and graphed on a Lineweaver-Burk plot.

Computational Analyses

The genome data of *M. smegmatis* mc^2 155 were collected from the primary annotation database (http://cmr.tigr.org/tigr-scripts/CMR/shared/Genomes.cgi) at The Institute for Genomic Research (TIGR). Protein databank searches were carried out at the BLAST server of the National Center for Biotechnology Information at the National Institutes of Health Bethesda, Md., USA (http://www.ncbi.nlm.nih.gov), the BLAST server of the Transport Classification Databank, TCDB (www.tcdb.org), and at TIGR. Multiple alignment analysis and graphical optimization were achieved with CLUSTALW (www.ebi.ac.uk/clustalw) and

BOXSHADE (www.ch.embnet.org/software/BOX_form.html). The phylogenetic tree was calculated with the neighbor joining method as implemented in CLUSTALW. The tree was graphically visualized by importing phylip (*.ph) files into the Treeview software available at http://taxonomy.zoology.gla.ac.uk/rod/treeview.html.

Acknowledgements

This study was supported by grants SFB473 to F.T., GK805 to E.F.P.-S., and SFB431 to K.J. of the Deutsche Forschungsgemeinschaft. Thanks to Michael Niederweis and Claudia Mailänder for many discussions and a gift of *M. smegmatis* DNA. We thank Lucille Schmieding and Manfred Beleut for editing the manuscript and Sébastien Rigali for critical reading.

References

Agarwal N, Tyagi AK: Mycobacterial transcriptional signals: requirements for recognition by RNA polymerase and optimal transcriptional activity. Nucleic Acids Res 2006;34: 4245–4257.

Angell S, Lewis CG, Buttner MJ, Bibb MJ: Glucose repression in *Streptomyces coelicolor* A3(2): a likely regulatory role for glucose kinase. Mol Gen Genet 1994;244:135–143.

Bercovier H, Vincent V: Mycobacterial infections in domestic and wild animals due to *Mycobacterium marinum, M. fortuitum, M. chelonae, M. porcinum, M. farcinogenes, M. smegmatis, M. scrofulaceum, M. xenopi, M. kansasii, M. simiae* and *M. genavense*. Rev Sci Tech 2001;20:265–290.

Deol P, Vohra R, Saini AK, Singh A, Chandra H, Chopra P, et al: Role of *Mycobacterium tuberculosis* Ser/Thr kinase PknF: implications in glucose transport and cell division. J Bacteriol 2005;187:3415–3420.

Hansen T, Reichstein B, Schmid R, Schönheit P: The first archaeal ATP-dependent glucokinase, from the hyperthermophilic crenarchaeon *Aeropyrum pernix*, represents a monomeric, extremely thermophilic ROK glucokinase with broad hexose specificity. J Bacteriol 2002;51:5955–5965.

Huang YS, Ge H, Zhang HM, Lei JQ, Zhang XL, Wang HH: Purification and characterization of the *Mycobacterium tuberculosis* FabD$_2$, a novel malonyl-CoA:AcpM transacetylase of fatty acid synthase. Protein Expr Purif 2006;45:393–399.

Imrisova K, Arreguin-Espinosa R, Guzman S, Rodriguez-Sanoja R, Langley E, Sanchez S: Biochemical characterization of the glucose kinase from *Streptomyces coelicolor* compared to *Streptomyces peucetius* var. *caesius*. Res Microbiol 2005;156:361–366.

Jayanthi Bai N, Ramachandra Pai M, Suryanarayana Murthy P, Venkitasubramanian TA: Pathways of carbohydrate metabolism in mycobacterium tuberculosis H37Rv1. Can J Microbiol 1975;21:1688–1691.

Kwakman JH, Postma PW: Glucose kinase has a regulatory role in carbon catabolite repression in *Streptomyces coelicolor*. J Bacteriol 1994;176:2694–2698.

Lunin VV, Li Y, Schrag JD, Ianuzzi P, Cygler M, Matte A: Crystal structures of *Escherichia coli* ATP-dependent glucokinase and its complex with glucose. J Bacteriol 2004;186: 6915–6927.

Mahr K, van Wezel GP, Svensson C, Krengel U, Bibb MJ, Titgemeyer F: Glucose kinase of *Streptomyces coelicolor* A3(2): large-scale purification and biochemical analysis. Antonie Van Leeuwenhoek 2000;78:253–261.

Mathur D, Ahsan Z, Tiwari M, Garg LC: Biochemical characterization of recombinant phosphoglucose isomerase of *Mycobacterium tuberculosis*. Biochem Biophys Res Commun 2005;337:626–632.

Mesak LR, Mesak FM, Dahl MK: *Bacillus subtilis* GlcK activity requires cysteines within a motif that discriminates microbial glucokinase into two lineages. BMC Microbiology 2004;3:4–6.

Parche S, Schmid R, Titgemeyer F: The phosphotransferase system of *Streptomyces coelicolor* identification and biochemical analysis of a histidine phosphocarrier protein HPr encoded by the gene *ptsH*. Eur J Biochem 1999;265:308–317.

Ramakrishnan T, Indira M, Maller RK: Evaluation of the routes of glucose utilization in virulent and avirulent strains of *Mycobacterium tuberculosis*. Biochim Biophys Acta 1962;59:529–532.

Ramos I, Guzman S, Escalante L, Imriskova I, Rodriguez-Sanoja R, Sanchez S, et al: Glucose kinase alone cannot be responsible for carbon source regulation in *Streptomyces peucetius* var. *caesius*. Res Microbiol 2004; 155:267–274.

Reed C, von Reyn CF, Chamblee S, Ellerbrock TV, Johnson JW, Trenschel RJ, et al: Environmental risk factors for infection with *Mycobacterium avium* complex. Am J Epidemiol 2006;2:32–40.

Sambrook J, Fritsch EF, Maniatis T: Molecular Cloning: A Laboratory Manual. Cold Spring Harbor/NY, Cold Spring Harbor Laboratory Press, 1989.

Sassetti CM, Boyd DH, Rubin EJ: Genes required for mycobacterial growth defined by high density mutagenesis. Mol Microbiol 2003; 48:77–84.

Sassetti CM, Rubin EJ: Genetic requirements for mycobacterial survival during infection. Proc Natl Acad Sci USA 2003;100:12989–12994.

Schiefner A, Gerber K, Seitz S, Welte W, Diederichs K, Boos W: The crystal structure of Mlc, a global regulator of sugar metabolism in *Escherichia coli*. J Biol Chem 2005;280: 29073–29079.

Sharma K, Gupta M, Pathak M, Gupta N, Koul A, Sarangi S, et al: Transcriptional control of the mycobacterial *embCAB* operon by PknH through a regulatory protein, EmbR, in vivo. J Bacteriol 2006;188:2936–2944.

Shawver LK, Slamon D, Ullrich A: Smart drugs: tyrosine kinase inhibitors in cancer therapy. Cancer Cell 2002;1:117–123.

Skarlatos P, Dahl MK: The glucose kinase of *Bacillus subtilis*. J Bacteriol 1998;180:3222–3226.

Smith MW, Neidhardt FC: Proteins induced by anaerobiosis in *Escherichia coli*. J Bacteriol 1983;154:336–343.

Titgemeyer F, Reizer J, Reizer A, Saier MH Jr: Evolutionary relationships between sugar kinases and transcriptional repressors in bacteria. Microbiology 1994a;140:2349–2354.

Titgemeyer F, Walkenhorst J, Cui X, Reizer J, Saier MH Jr: Proteins of the phosphoenolpyruvate:sugar phosphotransferase system in *Streptomyces*: possible involvement in the regulation of antibiotic production. Res Microbiol 1994b;145:89–92.

Trejo AG, Haddock JW, Chittenden GJ, Baddiley J: The biosynthesis of galactofuranosyl residues in galactocarolose. Biochem J 1971;122: 49–57.

Van Wezel GP, König M, Mahr K, Nothaft H, Thomae AW, Bibb M, Titgemeyer F: A new piece of an old jigsaw: glucose kinase is activated posttranslationally in a glucose transport-dependent manner in *Streptomyces coelicolor* A3(2). J Mol Microbiol Biotechnol 2007;12:65–72.

Waagmeester A, Thompson J, Reyrat JM: Identifying sigma factors in *Mycobacterium smegmatis* by comparative genomic analysis. Trends Microbiol 2005;13:505–509.

Wang R, Prince JT, Marcotte EM: Mass spectrometry of the M. smegmatis proteome: protein expression levels correlate with function, operons, and codon bias. Genome Res 2005;15:1118–1126.

Zeppenfeld T, Larisch C, Lengeler JW, Jahreis K: Glucose transporter mutants of *Escherichia coli* K-12 with changes in substrate recognition of IICB^{Glc} and induction behavior of the *ptsG* gene. J Bacteriol 2000;182:4443–4452.

Sugar Transport Systems of *Bifidobacterium longum* NCC2705

Stephan Parche[a] Johannes Amon[b] Ivana Jankovic[a] Enea Rezzonico[a]
Manfred Beleut[b] Hande Barutçu[b] Inke Schendel[b] Mike P. Eddy[c]
Andreas Burkovski[b] Fabrizio Arigoni[a] Fritz Titgemeyer[b]

[a]Nestlé Research Center, Vers-chez-les-Blanc, Lausanne, Switzerland; [b]Department of Microbiology, Friedrich Alexander University Erlangen-Nürnberg, Erlangen, Germany; [c]BioTechHome.com, Del Mar, Calif., USA

Key Words
Probiotics · Bifidobacterium · ABC family · Food microbiology · Nutrition

Abstract
Here we present the complement of the carbohydrate uptake systems of the strictly anaerobic probiotic *Bifidobacterium longum* NCC2705. The genome analysis of this bacterium predicts that it has 19 permeases for the uptake of diverse carbohydrates. The majority belongs to the ATP-binding cassette transporter family with 13 systems identified. Among them are permeases for lactose, maltose, raffinose, and fructooligosaccharides, a commonly used prebiotic additive. We found genes that encode a complete phosphotransferase system (PTS) and genes for three permeases of the major facilitator superfamily. These systems could serve for the import of glucose, galactose, lactose, and sucrose. Growth analysis of NCC2705 cells combined with biochemical characterization and microarray data showed that the predicted substrates are consumed and that the corresponding transport and catabolic genes are expressed. Biochemical analysis of the PTS, in which proteins are central in regulation of carbon metabolism in many bacteria, revealed that *B. longum* has a glucose-specific PTS, while two other species (*Bifidobacterium lactis* and *Bifidobacterium bifidum*) have a fructose-6-phosphate-forming fructose-PTS instead. It became obvious that most carbohydrate systems are closely related to those from other actinomycetes, with a few exceptions. We hope that this report on *B. longum* carbohydrate transporter systems will serve as a guide for further in-depth analyses on the nutritional lifestyle of this beneficial bacterium.

Introduction

The past decade of molecular biology research was dominated by the era of genome sequencing. Today, there are several hundreds of bacterial genomes available that need profound and detailed analysis. The body of data is so immense that such a challenge can only be approached by sophisticated bioinformatical analysis. Thus, genome sequencing and subsequent in silico analysis can provide a thorough understanding about a living organism for which no or little experimental data exist. While the initial publication of a genome [Bentley et al., 2002; Schell et al., 2002] can provide a road map, more in-depth work is required to identify sets of genes and describe distinct pathways such as stress response, nitrogen or carbon metabolism or regulation [Amar et al., 2002; Bertram et al., 2004].

Bifidobacteria belong to the group of high-GC Gram-positive actinomycetes. They are strictly anaerobic mi-

Table 1. Growth capacity of *B. longum* NCC2705 on 23 carbohydrates

Carbon source	Growth[a]	Predicted/validated system[b]
Monocarbohydrates		
Arabinose	++	nf
Fructose	+	12
Galactose	+++	17
Glucose	+++	14, 17
Glycerol	−	na
Mannose	−	na
Mannitol	−	na
N-acetylglucosamine	−	na
Rhamnose	−	na
Ribose	+	1
Sorbitol	−	na
Xylitol	−	na
D-Xylose	++	13
Di-/trisaccharides		
Maltose	++	3
Melibiose	++	4, 18
Salicine	−	na
Sucrose	++	15
Trehalose	−	na
Lactose	+++	5, 8, 16
Raffinose	+++	10
Oligosaccharides		
Arabinogalactan	+	nf
Raftilose (oligofructose)[c]	++	1, 3, 5, 6, 8, 11, 12, 15, 16
Starch	−	na

na = Not applicable; nf = not found.
[a] Growth capacity was scored as: slow growth (+), moderate growth (++) and fast growth (+++).
[b] Numbers refer to numbering of the carbohydrate systems as listed in table 2.
[c] Composition of raftilose was 93.2% oligofructose and up to 6.8% glucose, fructose, and sucrose.

croorganisms that live in the mammalian gastrointestinal tract, where they represent about 95% of the gut flora of breastfed babies and make up to 3% of the gut flora in adult humans [Biavati et al., 2001]. Together with lactobacilli, bifidobacteria are considered health-promoting bacteria and thus are used in food products [Abbott, 2004; Gibson et al., 1995b].

Bifidobacteria can grow on a wide range of carbon sources. Some of these, like oligofructose, inulin, and raffinose, have been identified as bifidogenic compounds and are used as food additives (prebiotics) to selectively promote growth of bifidobacteria in the gut [Gibson et al., 1995a]. Knowledge on molecular systems that par-ticipate in metabolizing these prebiotics is of primary interest for the understanding of the lifestyle of bifidobacteria. Furthermore, the understanding of the precise molecular mechanisms that underlie the positive effects of probiotics is considered to be the key towards the development of health-promoting food products. The publication of the genome sequence from *Bifidobacterium longum* NCC2705 revealed that the chromosome encodes numerous genes for carbohydrate utilization [Schell et al., 2002]. The latter publication also provides information that *B. longum* is equipped with more than 40 glycosyl hydrolases that are predicted to be involved in the degradation of higher order oligosaccharides.

In this communication, we compiled all carbohydrate transport systems that are present in *B. longum*. We then used microarray technology in combination with growth analysis and biochemical characterization of transport systems to evaluate the accuracy of our predictions. Hence, our investigation should serve as a guide for the molecular characterization of all carbohydrate transport systems present in *B. longum*.

Results and Discussion

Carbohydrate Growth Profile

We established a profile for carbon source utilization of *B. longum* NCC2705 using a defined, semisynthetic medium supplemented with the source of carbon to be monitored. As listed in table 1, the strain was able to utilize 13 out of 23 tested carbohydrates, among which were glucose, lactose, maltose, raffinose and oligofructose. Surprisingly, *B. longum* was not able to ferment the very common carbon sources glycerol and N-acetyglucosamine and other 'classical' carbon sources like mannitol that are metabolized by many bacteria. Interestingly, the profile shows a preference for di-, tri- and oligosaccharides, pointing towards a biased utilization for complex oligosaccharides that are found for example in human milk [Bode, 2006; Schell et al., 2002].

Prediction of Carbohydrate Transport Systems

We began the in silico analysis by conducting BLASTP analyses at the BLAST server of the TIGR institute (http://tigrblast.tigr.org/cmr-blast/) following previously described strategies [Bertram et al., 2004]. The detected gene products from *B. longum* NCC2705 and those of the adjacent genes were then searched with BLASTP at various genome servers, including the one of the Transporter Classification Database (TCDB; www.tcdb.org), in order

Table 2. Carbohydrate transport systems of *B. longum* NCC2705

	Family/substrate	BL[a]	Gene loci	Microarrays						Homologue/reference
				BL[a]	lac	raf	mal	FOS	CGH	
	ABC									
1	ribose	0033-0036	*rbsEKFG*	0033	0	0	0	+	7/9	*rbs* operon *E. coli* [Bell et al., 1986]
				0035	0	0	0	+	9/9	
				0036	0	0	0	+	9/9	
2	disaccharide	0055	*abcG*	0055	nd	0	nd	nd	7/9	distantly related to disaccharide ABC permeases
3	maltose	0141-0146	*malERFG*	0141	0	0	+	−	9/9	*malF E. coli* [Froshauer et al., 1988; Schell et al., 2002]
			urf sugH	0143	0	0	+	0	9/9	
				0144	0	0	+	0	9/9	
				0145	0	0	+	0	9/9	
				0146	0	0	+	+	6/9	
4	unknown	0176-0177, 0188-0190	*agaRA* *abcEFG*	0188	0	0	0	0	6/9	distantly related to various ABC systems [Schell et al., 2002]
				0189	0	0	0	0	6/9	
				0190	0	0	0	0	6/9	
5	lactose, FOS	0258-0264	*lacR bga* *lacGFE sugK*	0259	0	0	0	0	9/9	*lacEFG S. coelicolor* [Bertram et al., 2004]; *bga* from *B. stearothermophilus* [Hirata et al., 1986; Schell et al., 2002]
				0260	0	0	0	0	5/9	
				0261	0	0	nd	0	3/9	
				0262	0	0	0	+	2/9	
6	FOS	0423-0426	*oliGFER*	0423	0	0	0	0	6/9	distantly related to disaccharide ABC permeases [Schell et al., 2002]
				0424	nd	nd	nd	nd	nd	
				0425	0	0	0	+	6/9	
				0426	0	0	0	0	6/9	
7	α-glucosides	0523-0525	*agl abcGR*	0524	0	0	0	0	7/9	*ngcG S. coelicolor* [Bertram et al., 2004]
				0523	0	0	nd	nd	7/9	
8	lactose, FOS	1163-1171	*msiE1E2E3 abfA2* *msiR1 bga msiGFR2*	1163	0	0	0	0	8/9	*lac* operon *S. coelicolor* [Bertram et al., 2004; Schell et al., 2002]
				1164	+	0	0	+	7/9	
				1165	+	0	nd	+	6/9	
				1169	0	0	nd	0	6/9	
				1170	+	0	0	0	7/9	
9	mannose-like oligosaccharides	1327-1332	*amaABCEFG*	1327	+	0	−	−	6/9	SCO0948-0952 of *S. coelicolor* [Bertram et al., 2004; Schell et al., 2002]
				1328	0	0	nd	0	6/9	
				1329	0	0	0	0	1/9	
				1330	0	0	0	0	1/9	
				1331	0	0	nd	0	1/9	
				1332	0	0	0	0	1/9	
10	raffinose	1518-1526	*aga rafREFG* *urf agl*	1521	0	+	nd	0	9/9	distantly related to carbohydrate ABC permeases [Schell et al., 2002]
				1522	nd	+	nd	nd	9/9	
				1523	0	+	nd	0	9/9	
11	FOS	1638-1640	*fosEFG*	1638	0	0	0	+	3/9	distantly related ABC carbohydrate transporter
				1639	0	0	0	+	5/9	
				1640	0	0	0	+	nd	
12	multiple sugar, fructose, mannose	1691-1696	*glkA1 fmaKREFG*	1694	nd	nd	nd	+	7/9	fructose-ABC permease from *A. radiobacter* [Williams et al., 1995]
				1695	nd	nd	nd	nd	7/9	
				1696	0	+	nd	+	7/9	
13	xylosides, xylose	1704-1710	*xylAPGXBR*	1706	0	nd	nd	0	6/9	[Bertram et al., 2004]
	PTS									
14*	glucose	0411-0412, 1632-1634	*ptsHI ptsG licT glkA2*	0411	0	0	0	0	7/9	[Brückner et al., 2002; Parche et al., 2001]
				0412	0	0	0	0	9/9	
				1632	0	nd	0	nd	7/9	

Table 2 (continued)

Family/substrate	BL[a]	Gene loci	Microarrays						Homologue/reference
			BL[a]	lac	raf	mal	FOS	CGH	
MFS									
15 sucrose	0105-0107	cscABR	0105	0	nd	nd	+	9/9	cscB E. coli [Bockmann et al., 1992]
			0106	0	0	+	+	8/9	
			0107	0	0	0	+	9/9	
16 lactose	0976-0978	lacS urf lacZ	0976	0	0	0	0	7/9	lacS S. thermophilus [Veenhoff et al., 2001]
			0978	+	0	+	+	7/9	
17* glucose glalactose	1630-1631	pgm glcP	1631	–	–	–	–	7/9	glcP S. coelicolor [van Wezel et al., 2005]
GPH									
18 unknown	0165	blo0165	0165	0	0	nd	0	7/9	distantly related to Na+/melibiose and pentoside permeases
MIP									
19 glycerol-like	0410	glpF	0410	0	0	0	–	9/9	glpF E. coli [Hindle et al., 1994]

The given references were selected to provide information for representative, well-studied homologues.
nd = Not detectable; 0 = expressed; + = induced; – = repressed; CGH = comparative genomic hybridization of B. longum biotypes.
* An experimental validated system.
[a] Protein annotation number of the B. longum genome (www.tigr.org).

to identify all carbohydrate transporter genes [Saier et al., 2006]. As depicted in table 2 and figure 1, our search led to the identification of 19 B. longum loci predicted to encode carbohydrate transporters. They belong to the ATP-binding cassette family (ABC), the phosphotransferase system (PTS), the major facilitator superfamily (MFS), the glycoside-pentoside-hexuronide cation symporter family (GPH), and the major intrinsic protein family (MIP). The majority was formed by 13 ABC permeases (ATP-binding cassette) [Bertram et al., 2004], which include porters for di-, tri- and higher order oligosaccharides. Three permeases of the MFS, which include proton symporters and facilitators, are predicted to transport glucose, lactose, and sucrose [Pao et al., 1998]. We also found one complete glucose-specific PTS. It comprises of the general phosphotransferases enzyme I (EI) and HPr, which operate in global carbon regulation in many bacteria [Brückner et al., 2002], and a single predicted glucose-specific enzyme II permease. Finally, a putative glycerol permease of the MIP and one permease of the GPH family completed the list of detected NCC2705 sugar transport systems.

The genetic organization of all genes encoding the transport systems is summarized in figure 2. To gain more hints on possible substrates, we inspected the neighborhood for associated metabolic genes, which are usually co-localized in an operon structure, and for possible regulators, which often occur adjacent to the target operon.

ABC Family: Lactose, Raffinose, Maltose, and Fructooligosaccharides Transport

We performed microarray experiments with RNA isolated from B. longum cultures grown on lactose, maltose, raffinose, and fructooligosaccharides (FOS) to evaluate the accuracy of our predictions (table 2). Array data that were derived from RNA samples prepared from glucose-grown cultures served as the reference to calculate the relative expression. We found that almost all genes encoding the components for sugar transport and utilization were expressed. In addition, the conservation of the identified genes was demonstrated by comparative genome hybridizations in another five B. longum biotype *longum*, one B. longum biotype *suis*, and two B. longum biotype *infantis*. As indicated in table 2, the vast majority of these genes was conserved in these closely related bifidobacteria strains. Only the predicted ABC permease encoded by BL1329-1332 was found to be specific for B. longum NCC2705.

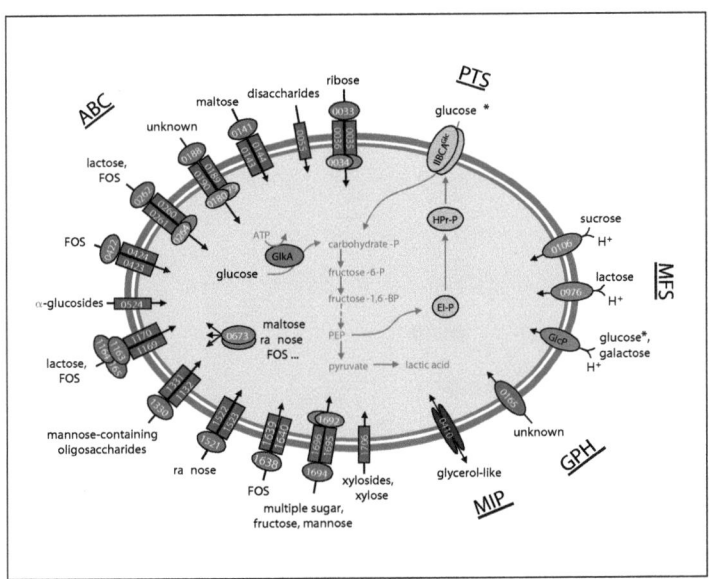

Fig. 1. Sugar transport systems of *B. longum* NCC2705. Shown are the permeases of the ABC (red), PTS (yellow), MFS (green), MIP (deep purple) and GPH family (blue). The derived putative substrates are inferred from in silico analyses in combination with experimental data. Experimentally verified systems are denoted by an asterisk.

Lactose Transport. In our analysis, we predicted three potential lactose permeases, two of the ABC-type and one of the MFS-type (see below) (table 2). The first ABC permease encoded by *bl0260-0262* is similar to the predicted lactose ABC permease of *Streptomyces coelicolor* [Bertram et al., 2004] and was expressed under all conditions tested. Additional evidence that this operon encodes a lactose permease is brought by the presence within the same transcriptional unit of a β-galactosidase gene related to those found in bacilli [Hirata et al., 1986]. It should be noted that the gene for the substrate-binding protein (BL0262) is induced by FOS, which would indicate that another sugar besides lactose is recognized. The second lactose transporter is encoded by *bl1163-1165* and *bl1169-1170*. This gene locus shows a feature that has not been described previously, as there are three consecutive genes encoding for extracellular substrate-binding proteins. The gene for the membrane protein BL1170 was induced by lactose. Furthermore, two of the genes that encode substrate-binding proteins (*bl1164* and *bl1165*) were induced by lactose and FOS (table 1). A slight difference in induction suggests that BL1164 might be a lactose-binding protein and BL1165 a fructose or FOS-binding protein. The presence of three genes coding for sugar-binding domains as well as their induction by at least two different sugars indicates that this operon might encode a permease with multiple sugar transport function.

Raffinose Transport. DNA microarray experiments performed with RNA from *B. longum* cells grown in the presence of the trisaccharide raffinose showed an enhanced expression of the transporter genes *bl1521-1523* (*rafEFG*) and *bl1696* (*fmaG*, see below). *rafEFG* encodes

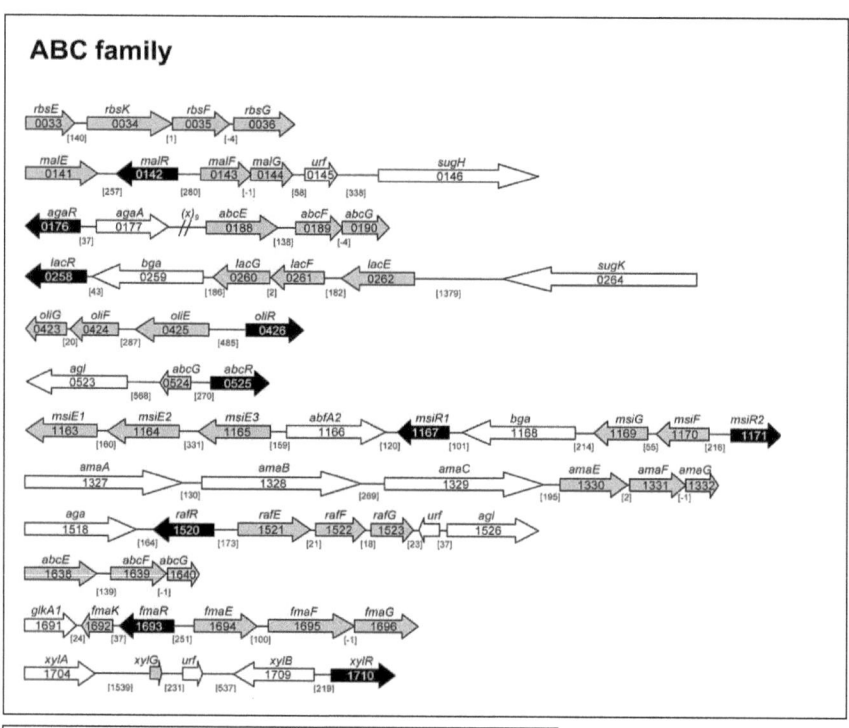

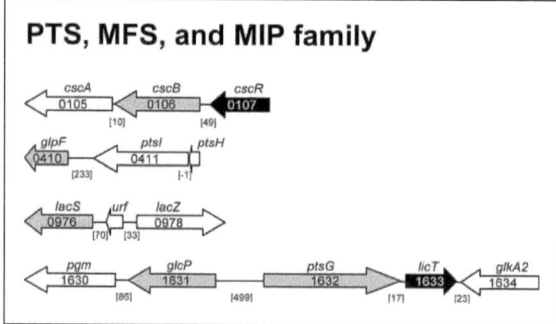

Fig. 2. Location of permease genes in the chromosome of *B. longum*. The arrows indicate the length and transcriptional orientation of annotated genes and predicted open reading frames. Genes encoding transport systems are depicted in gray and regulator genes are highlighted in black. Genes are denominated by their number with the prefix 'bl'. The gene names are assigned according to the annotations given by The Institute for Genomic Research (http://www.tigr.org) and by us. Monocistronic operons are not shown. *urf*, unknown reading frame. (x)$_9$, space of nine *orfs* in between. For further details see text.

an ABC permease with distant similarity to other ABC carbohydrate transporters. The *raf* operon is preceded by a ROK family regulator gene [Titgemeyer et al., 1994] and a gene for an α-galactosidase *(aga, bl1518)*, which would cleave raffinose into galactose and sucrose (fig. 2). In addition, a homologue of an α-glucosidase *(agl, bl1526)* is situated two genes further downstream (fig. 2). Agl, however, is unlikely to cleave sucrose, which is a 1,2-β-D-fructofuranose. A sucrose hydrolase could be encoded by *cscA (bl0105,* table 2), which is located elsewhere on the chromosome. Therefore, we suggest that the ABC permease RafEFG transports probably other oligosaccharides besides raffinose.

Maltose Transport. Growth of *B. longum* NCC2705 on maltose led to the induction of the gene locus *bl0141-0144 (malERFG),* which is in line with in silico predicting these genes to constitute a maltose transport system. Interestingly, the MalEFG proteins show higher similarity to the *Escherichia coli* [Froshauer et al., 1988] maltose transporter than to MalEFG of the closer related *S. coelicolor* [Bertram et al., 2004], thus suggesting a different evolutionary origin. The two downstream genes *bl0145-0146* are also induced by maltose, suggesting a functional association to the *mal* operon. While *bl0145* is a gene of unknown function, the product of *bl0146 (sugH,* sugar hydrolase) has low similarity to an arabinosidase from *Butyrivibrio fibrisolvens* and a xylanase from *Streptomyces avermitilis*. *sugH* is also induced by FOS, suggesting that its product is involved in different metabolic pathways.

FOS Transport. Raftilose is a commonly used prebiotic composed of 93.2% FOS and up to 6.8% glucose, fructose, and sucrose. Growth on raftilose led to the induction of nine transport systems (table 2). The first in the table codes for an operon *(bl0033-0036, rbsEKFG)* whose gene products display similarity to the *E. coli* ribose ABC-porter. Interestingly, these *B. longum* proteins are more similar to those encoded by the *ytfQRT yjfF E. coli* operon of unknown function. We suggest for this porter that it may transport ribose or ribose-containing oligosaccharides. However, the observation that the *rbs* genes are induced by FOS would argue against this suggestion. Hence, experimental investigation is required to clarify the substrate specificity of this ABC-permease. While the FOS-inducible genes *bl0146, bl0262,* and *bl1164-1165* are already discussed above, there were another three FOS-inducible ABC permease operons, *bl0423-0425 (oliGFER,* oligosaccharide), *bl1638-1640 (fosEFG),* and *bl1694-1696 (fmaEFG)*. The first two systems are distantly related to other ABC carbohydrate transport systems and thus were predicted as putative porters for oli-gofructose. The third system is another ABC transporter *(fma,* fructose-mannose-like) with similarity to a fructose ABC permease of *Agrobacterium radiobacter* [Williams et al., 1995]. A glucose kinase gene *(glkA),* a gene for an ATPase *(fmaK)* and a ROK family regulator are found upstream of *fmaEFG*. Hence, we predict that this porter may transport multiple sugars with fructose, glucose and mannose moieties.

Other Oligosaccharide Transporters. The gene loci *bl0055 (abcG), bl0523-0525 (agl abcGR)* and *bl1327-1332 (amaABCEFG)* encode ABC permease genes that are distantly related to other ABC carbohydrate transporters. The first two systems only contain a single gene for one ABC permease membrane protein, but no gene coding for a substrate-binding protein. The C-terminal sequence of BL0524 is related to NgcG (predicted chitobiose permease) [Bertram et al., 2004], while the N-terminal half shows no similarity to any protein in the databases. Hence, it remains to be seen whether this unique gene encodes a carbohydrate permease. However, since *bl0524* is preceded by a predicted α-glucosidase, the permease could be a porter for higher order oligosaccharides containing α-glucosides. The gene locus *bl1327-1332* is particularly interesting as it only occurs in *B. longum* NCC2705 (table 2). The permease genes are preceded by three paralogous genes, whose products exhibit 38–42% identity to the putative α-mannosidase SCO0948 of *S. coelicolor* [Bentley et al., 2002]. Furthermore, the gene order of the *B. longum* loci is found conserved to the gene locus *sco0948-0952* of *S. coelicolor*, which encodes a predicted disaccharide/oligosaccharide ABC permease [Bertram et al., 2004]. We suggest that this permease transports mannose-containing oligosaccharides.

Xylose Transport. Gene locus *bl1704-1710* comprises of a gene that encodes a truncated, possibly not functional ABC permease *(xylG),* the xylose metabolic genes *xylA* (xylose isomerase) and *xylB* (xylulose kinase), and a ROK family regulator gene *xylR*. XylA and XylB share 43 and 30% protein identity to the respective *E. coli* proteins [Lawlis et al., 1984]. Hence, this locus most likely encodes an inducible regulon for xylose metabolism. It is however unclear whether the truncated XylG permease contributes to xylose transport.

Carbohydrate-Specific ABC ATPases. The analysis of ABC operons revealed that only 5 of the 13 systems included a gene for a required ATPase (table 2; *bl0034, bl0264, bl0179, bl0180, bl1692*). This situation is comparable to what we have previously found in *S. coelicolor* A3(2) [Bertram et al., 2004]. This bacterium has a global multiple sugar import protein MsiK that assists those ABC systems

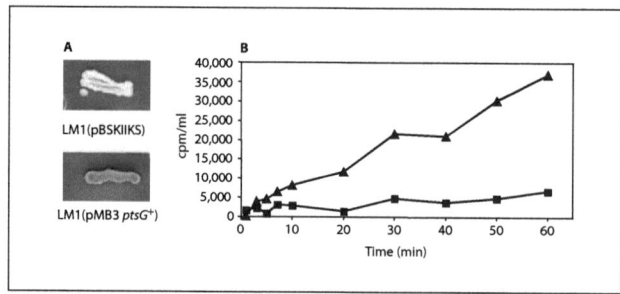

Fig. 3. Functional complementation of PtsG[Blo] and activity assay. **A** The figure shows a MacConkey glucose agar plate with LM1(pBSKIIKS, control) and LM1(pMB3 ptsG+). Formation of red colonies (dark gray) indicate restoration of glucose fermentation, while white colonies indicated a glucose-negative phenotype (light gray). **B** A PTS assay is shown with PtsG[Blo]-containing membranes prepared from LM1(pMB3 ptsG+) (triangles) and with control membranes prepared from LM1(pBSKIIKS) (squares). PEP-dependent phosphorylation of glucose was monitored in the presence of purified S. coelicolor EI and HPr over a range of 60 min. Only membrane vesicles with PtsG[Blo] showed PTS-specific phosphorylation.

lacking their own ATPase. In BL0673, we found a homologue with 61% protein identity to MsiK[Sco] and predict that bl0673 encodes a universal ATPase for the B. longum NCC2705 ABC-type carbohydrate transporters.

Glucose Transport and Phosphotransferase System

Unlike many other bacteria, B. longum preferentially uses lactose rather than glucose as the primary carbon source [Parche et al., 2006]. We showed by microarray analysis that this is achieved in part by transcriptional repression of only one gene, glcP (bl1631). We demonstrated that glcP encodes an inducible glucose-specific proton symporter of the MFS (table 2), which is responsible for the uptake of glucose. Substrate competition assays revealed that the permease could efficiently recognize galactose. Thus, we wish to suggest that GlcP is also the permease for this sugar (table 1). Our BLASTP analysis revealed that glcP is present in Bifidobacterium adolescentis sharing 83% identity at the protein level. Interestingly, the genetic context is different in the two bifidobacteria with respect to glcP. While B. longum has the gene order pgm-glcP-ptsG-licT-glkA2, the same region in B. adolescentis is lacking glcP, which is located somewhere else on the chromosome.

In addition, the B. longum NCC2705 genome exhibits a predicted glucose-specific PTS (bl0411-0412, bl1632; ptsHI, ptsG). Our gene expression analysis indicates that these genes are constitutively expressed albeit at low level, suggesting that, under the conditions tested, the system does not contribute to glucose intake. Nevertheless, heterologous expression of B. longum PtsG in E. coli LM1, lacking all glucose transport systems, led to the restoration of glucose fermentation (fig. 3A). The activity of PtsG[Blo] was further demonstrated in a PTS assay. We therefore prepared PtsG[Blo]-containing membrane vesicles from LM1(pMB3 ptsG+), added purified EI and HPr from S. coelicolor and measured phosphoenolpyruvate (PEP)-dependent phosphorylation of radiolabeled glucose (fig. 3B). A significant increase of glucose-phosphate formation was observed, while in a control assay vesicles from LM1(pBSKS) without PtsG[Blo] did not give significant formation of glucose-6-phosphate.

We then examined experimentally whether such a PTS activity could be found in representatives of two other species, namely Bifidobacterium bifidum MB245 and Bifidobacterium lactis DSM10140. Surprisingly, we could not detect any glucose-PTS activity. Instead, we found that both strains possess a fructose-specific PTS. This was demonstrated by a series of transphosphorylation assays. The assay is based on the idea that enzyme II permeases have two distinct substrate-binding sites, one for the sugar and one for the sugar phosphate, and the phosphoryl group can be exchanged between the two substrates [Saier et al., 1977]. We found that in membrane vesicles from cells grown in complex medium supplemented with fructose, a phosphoryl group exchange occurred from non-radioactive fructose-6-phosphate (phosphoryl group donor) to radiolabeled fructose (phosphoryl acceptor). The specific activity was 1.6 ± 0.5 nmol fructose-6-phosphate per mg protein per minute in *B. bifidum* and 1.3 ± 0.4 U in *B. lactis*. No transphosphorylation (<0.1 U) was observed when fructose-1-phosphate was applied as the phosphoryl group donor. Altogether, these data show that the two strains have a fructose-6-phosphate-forming PTS. However, it is unusual since most described fructose-PTSs deliver fructose-1-phosphate [Nothaft et al., 2003]. Here, the detected PTS-type would make sense as bifidobacteria funnel carbohydrates via fructose-6-phosphate into the so-called bifidus shunt.

The role of the PTS in bifidobacteria needs further molecular analysis. Our data suggest that some species have a glucose-PTS, while others have a fructose-PTS.

MFS Family: Lactose, Sucrose, and Glucose Transport

While most transport systems of *B. longum* are related to those from other actinomycetes, in particular streptomycetes, we found three systems which are most similar to lactic acid bacteria and to enteric bacteria. The amino acid sequences of the putative sucrose permease CscB (BL0106) of the MFS and the adjacent sucrose hydrolase CscA are homologous to proteins composing a system that has been described by Bockmann et al. [1992] in an *E. coli* isolate. The *csc* genes of *B. longum* were induced by raftilose, which also contains a significant portion of sucrose (table 2). Another MFS-type protein is the putative lactose permease LacS (BL0976) that shared 55% protein identity to LacS⁻ from *Lactobacillus plantarum* WCFS1 [Kleerebezem et al., 2003] and 48% to the well-characterized LacS of *Streptococcus thermophilus* [Veenhoff et al., 2001]. While LacS[Sth] has an additional enzyme IIA-domain at the C-terminus that is involved in a sophisticated carbon regulatory pathway [van den Bogaard et al., 2000], LacS[Blo] is lacking this C-terminal extension. The third MFS system is the glucose/galactose transporter GlcP, which we have described in the preceding chapter above.

MIP- and GPH-Type Transporters

Analysis of the *B. longum* NCC2705 genome revealed that the bacterium possesses one permease from the MIP and one from the GPH family. The MIP member is predicted as a glycerol facilitator. Surprisingly, *B. longum* has no obvious glycerol kinase gene nor can it grow on glycerol. Hence, this facilitator may serve for the uptake of a glycerol-like carbohydrate. The GPH member has very distant similarity to melibiose and pentoside permeases and is apparently transcribed as a monocistronic operon.

Conclusions

Our combined analysis of the carbohydrate transporter systems of *B. longum* NCC2705 revealed 19 transport systems and provided a number of interesting conclusions: (1) *B. longum* can transport a variety of disaccharides and oligosaccharides like oligofructose, which are described as growth-promoting prebiotics. Compared to the initial publication of the genome sequence [Schell et al., 2002], in which 7 oligosaccharide systems were presented, our more extensive analysis revealed that up to 15 such systems may exist in NCC2705. (2) More than half of the transport systems are ATP-dependent ABC-type permeases, a feature that is also found in other actinomycetes such as *S. coelicolor* [Bertram et al., 2004] and *Mycobacterium smegmatis* [pers. unpubl. data]. (3) PTSs are present in bifidobacteria and occur either as glucose- or fructose-specific PTS. It remains to be established whether these PTS components serve additional regulatory function besides sugar transport as they do in other bacteria [Brückner et al., 2002; Rigali et al., 2006]. (4) Our comparative genome analysis with DNA microarrays revealed that most genes for carbohydrate permeases from *B. longum* are found in other closely related *Bifidobacterium* strains. This predictive study on the carbohydrate transporters from *B. longum* will hopefully facilitate and stimulate further detailed analysis that is required to unravel the nutritional lifestyle of bifidobacteria.

Experimental Procedures

Bacterial Strains, Growth Conditions and Plasmids

B. longum NCC2705 was isolated from human infant feces [Schell et al., 2002]. *B. lactis* DSM10140 and *B. bifidum* MB245 were kindly provided by Matthias Ehrmann (Technische Universität München). Cells of *B. longum* were grown anaerobically at 37°C either as a static culture using the Gas-pack system (AnaeroGen™, Oxoid) or with stirring (200 rpm) in a 0.5 l fermenter (Sixfors, INFORS) under a CO_2 atmosphere. Bacterial growth on different carbon sources was evaluated in three steps. In the first phase, sugar utilization was screened using API strips (BioMérieux, Germany). In the next step, growth was evaluated in static cultures on semisynthetic growth medium [Perrin et al., 2001] supplemented with 5 mM L-cysteine, by measuring the final optical density at OD_{600} after 48 h. Finally, growth curves were performed on some selected sugars by monitoring kinetics of OD_{600} of the cells grown in fermenter. *ptsG (bl1632)* was amplified by polymerase chain reaction using primers CACAG<u>AAGCTT</u>AC-GGTAGCAAGGAGGAACCG and ACGC<u>GGATCC</u>AACGTC-ATTGTTCCACATCG (restriction sites HindIII and BamHI are underlined). The fragment was cloned into pBluescript II KS(+) (Stratagene). The insert was confirmed by DNA sequencing and the resulting plasmid was designated pMB3. The *E. coli* strain LM1 [van Wezel et al., 2005] that is deficient in all glucose uptake systems was transformed with either pMB3 or pBSKIIKS(+). Resulting transformant colonies were streaked onto McConkey agar plates supplemented with 50 mM glucose to monitor glucose fermentation.

Microarray-Based Gene Expression Analysis

B. longum was grown in fermenter containing semisynthetic growth medium supplemented with 2% glucose, lactose, raffinose, maltose, and Raftilose P95 (FOS; Orafti), respectively. 15-ml culture samples were taken during exponential growth phase at OD_{600} of 0.5 to prepare total RNA. RNA preparation and microarrays were essentially performed as described [Parche et al., 2006]. RNA quality was controlled using the Agilent 2100 Bioanalyzer by looking at the integrity of 16S and 23S rRNAs.

Glass microarrays were produced by Eurogentec SA (Belgium) by printing PCR-based probes designed to cover approximately 97% of the identified ORFs of the *B. longum* NCC2705 genome. Labeling of RNA, cDNA synthesis and hybridizations were performed using the 3DNA Array 350RP Genisphere kit (Genisphere Inc., Hatfield, Pa., USA), following the protocol provided by the supplier. Relative gene expression ratios were then calculated using expression of glucose-grown cells as a reference. Mean relative gene expression values for each sugar represent the average of four hybridizations from two biological repeats (two hybridizations for each biological repeat) for the lactose versus glucose and the raffinose versus glucose comparisons, and two hybridizations from one biological repeat for the maltose versus glucose and the raftilose versus glucose comparisons. Genes were considered as differentially expressed if they displayed an average absolute $\log_2$-transformed gene expression ratio ≥ 2.

Comparative genome hybridization on microarrays was performed by using chromosomal DNA from nine *B. longum* ssp. Further details on the procedure are available as supplementary material.

PTS Assays

Transphosphorylation assays were performed as described [Saier et al., 1977; Titgemeyer et al., 1995]. The assay was conducted at 30°C in a 1-ml reaction volume containing 0.5–2 mg membrane vesicles, 10 mM non-radioactive sugar phosphate (phosphoryl group donor) and 10 μM radiolabeled [^{14}C]sugar (500,000 cpm per reaction). Samples were taken periodically up to 120 min. In vitro reconstitution of PTS-dependent phosphorylation was essentially conducted as described [Titgemeyer et al., 1995]. In a 100-μl reaction volume, 150 μg of PtsG-containing membrane vesicles, that were isolated from extracts of LM1(pMB3 ptsG$^+$), were combined with 5 mM PEP and 0.25 μg of purified histidine-tagged EI and 2 μg of pure histidine-tagged HPr. The reaction was initiated by addition of [^{14}C]glucose (100,000 cpm per reaction) at a final concentration of 200 μM. Samples were taken from 1 to 60 min to determine PEP-dependent phosphorylation of glucose by oscillographic scintillation counting. As a negative control membranes from LM1(pBSK), which did not contain PtsG, were measured under the same conditions in parallel. Data were derived from three independent experiments.

Computational Analyses

The genome data of *B. longum* NCC2705 were collected from the primary annotation database (http://cmr.tigr.org/tigr-scripts/CMR/shared/Genomes.cgi) at The Institute for Genomic Research (TIGR). Protein databank searches were carried out at the BLAST server of the National Center for Biotechnology Information at the National Institutes of Health Bethesda, Md., USA (http://www.ncbi.nlm.nih.gov), the BLAST server of the Transport Classification Databank, TCDB (www.tcdb.org), and at TIGR. Each identified protein was further analyzed by BLASTP using the genome server of *E. coli* (http://genolist.pasteur.fr/Colibri/), *B. subtilis* (http://genolist.pasteur.fr/SubtiList/), *Streptomyces avermitilis* (http://avermitilis.ls.kitasato-u.ac.jp/) and *S. coelicolor* (http://streptomyces.org.uk).

Acknowledgements

This study was supported through grants of SFB473 and GK805 of the Deutsche Forschungsgemeinschaft. We thank Michael Ehrmann and Bernd Eikmanns for strains and discussions.

References

Abbott A: Microbiology: gut reaction. Nature 2004;427:284–286.

Amar P, Ballet P, Barlovatz-Meimon G, Benecke A, Bernot G, Bouligand Y, et al: Hyperstructures, genome analysis and I-cells. Acta Biotheor 2002;50:357–373.

Bell AW, Buckel SD, Groarke JM, Hope JN, Kingsley DH, Hermodson MA: The nucleotide sequences of the *rbsD*, *rbsA*, and *rbsC* genes of *Escherichia coli* K12. J Biol Chem 1986;261:7652–7658.

Bentley SD, Chater KF, Cerdeno-Tarraga AM, Challis GL, Thomson NR, James KD, et al: Complete genome sequence of the model actinomycete *Streptomyces coelicolor* A3(2). Nature 2002;417:141–147.

Bertram R, Schlicht M, Mahr K, Nothaft H, Saier MH Jr, Titgemeyer F: *In silico* and transcriptional analysis of carbohydrate uptake systems of *Streptomyces coelicolor* A3(2). J Bacteriol 2004;186:1362–1373.

Biavati B, Mattarelli P: The family Bifidobacteriaceae; in Dworkin M, Falkow S, Rosenberg E, Schleifer KH, Stackebrandt E (eds): The Prokaryotes. New York, Springer, 2001, pp 1–70.

Bockmann J, Heuel H, Lengeler JW: Characterization of a chromosomally encoded, non-PTS metabolic pathway for sucrose utilization in *Escherichia coli* EC3132. Mol Gen Genet 1992;235:22–32.

Bode L: Recent advances on structure, metabolism, and function of human milk oligosaccharides. J Nutr 2006;136:2127–2130.

Brückner R, Titgemeyer F: Carbon catabolite repression in bacteria: choice of the carbon source and autoregulatory limitation of sugar utilization. FEMS Microbiol Lett 2002; 209:141–148.

Froshauer S, Green GN, Boyd D, McGovern K, Beckwith J: Genetic analysis of the membrane insertion and topology of MalF, a cytoplasmic membrane protein of *Escherichia coli*. J Mol Biol 1988;200:501–511.

Gibson GR, Beatty ER, Wang X, Cummings JH: Selective stimulation of bifidobacteria in the human colon by oligofructose and inulin. Gastroenterology 1995a;108:975–982.

Gibson GR, Roberfroid MB: Dietary modulation of the human colonic microbiota: introducing the concept of prebiotics. J Nutr 1995b;125:1401–1412.

Hindle Z, Smith CP: Substrate induction and catabolite repression of the *Streptomyces coelicolor* glycerol operon are mediated through the GylR protein. Mol Microbiol 1994;12: 737–745.

Hirata H, Fukazawa T, Negoro S, Okada H: Structure of a β-galactosidase gene of *Bacillus stearothermophilus*. J Bacteriol 1986;166: 722–727.

Kleerebezem M, Boekhorst J, van Kranenburg R, Molenaar D, Kuipers OP, Leer R, et al: Complete genome sequence of *Lactobacillus plantarum* WCFS1. Proc Natl Acad Sci USA 2003; 100:1990–1995.

Lawlis VB, Dennis MS, Chen EY, Smith DH, Henner DJ: Cloning and sequencing of the xylose isomerase and xylulose kinase genes of *Escherichia coli*. Appl Environ Microbiol 1984;47:15–21.

Nothaft H, Parche S, Kamionka A, Titgemeyer F: In vivo analysis of HPr reveals a fructose-specific phosphotransferase system that confers high-affinity uptake in *Streptomyces coelicolor*. J Bacteriol 2003;185:929–937.

Pao SS, Paulsen IT, Saier MH Jr: Major facilitator superfamily. Microbiol Mol Biol Rev 1998; 62:1–34.

Parche S, Thomae AW, Schlicht M, Titgemeyer F: *Corynebacterium diphtheriae*: a PTS view to the genome. J Mol Microbiol Biotechnol 2001;3:415–422.

Parche S, Beleut M, Rezzonico E, Jacobs D, Arigoni F, Titgemeyer F, et al: Lactose-over-glucose preference in *Bifidobacterium longum* NCC2705:glcP, encoding a glucose transporter, is subject to lactose repression. J Bacteriol 2006;188:1260–1265.

Perrin S, Warchol M, Grill JP, Schneider F: Fermentations of fructo-oligosaccharides and their components by *Bifidobacterium infantis* ATCC 15697 on batch culture in semi-synthetic medium. J Appl Microbiol 2001; 90:859–865.

Rigali S, Nothaft H, Noens EE, Schlicht M, Colson S, Müller M, et al: The sugar phosphotransferase system of *Streptomyces coelicolor* is regulated by the GntR-family regulator DasR and links N-acetylglucosamine metabolism to the control of development. Mol Microbiol 2006;61:1237–1251.

Saier MH Jr, Feucht BU, Mora WK: Sugar phosphate: sugar transphosphorylation and exchange group translocation catalyzed by the enzyme 11 complexes of the bacterial phosphoenolpyruvate: sugar phosphotransferase system. J Biol Chem 1977;252:8899–8907.

Saier MH Jr, Tran CV, Barabote RD: TCDB: the Transporter Classification Database for membrane transport protein analyses and information. Nucleic Acids Res 2006;34: D181–D186.

Schell MA, Karmirantzou M, Snel B, Vilanova D, Berger B, Pessi G, et al: The genome sequence of *Bifidobacterium longum* reflects its adaptation to the human gastrointestinal tract. Proc Natl Acad Sci USA 2002;99: 14422–14427.

Titgemeyer F, Reizer J, Reizer A, Saier MH Jr: Evolutionary relationships between sugar kinases and transcriptional repressors in bacteria. Microbiology 1994;140:2349–2354.

Titgemeyer F, Walkenhorst J, Reizer J, Stuiver MH, Cui X, Saier MH Jr: Identification and characterization of phosphoenolpyruvate: fructose phosphotransferase systems in three *Streptomyces* species. Microbiology 1995;141:51–58.

Van den Bogaard PT, Kleerebezem M, Kuipers OP, de Vos WM: Control of lactose transport, β-galactosidase activity, and glycolysis by CcpA in *Streptococcus thermophilus*: evidence for carbon catabolite repression by a non-phosphoenolpyruvate-dependent phosphotransferase system sugar. J Bacteriol 2000;182:5982–5989.

Van Wezel GP, Mahr K, König M, Traag BA, Pimentel-Schmitt EF, Willimek A, et al: GlcP constitutes the major glucose uptake system of *Streptomyces coelicolor* A3(2). Mol Microbiol 2005;55:624–636.

Veenhoff LM, Heuberger EH, Poolman B: The lactose transport protein is a cooperative dimer with two sugar translocation pathways. EMBO J 2001;20:3056–3062.

Williams SG, Greenwood JA, Jones CW: *Agrobacterium radiobacter* and related organisms take up fructose via a binding-protein-dependent active-transport system. Microbiology 1995;141:2601–2610.

From: Corynebacteria: Genomics and Molecular Biology. Edited by: Andreas Burkovski.

General and Regulatory Proteolysis in Corynebacteria 13

Johannes Amon, Alja Lüdke, Fritz Titgemeyer and Andreas Burkovski

Abstract
Proteases comprise a broad variety of functions in cells. They are involved in the degradation of polypeptides to supply amino acids for the synthesis of new proteins or for catabolism as carbon and energy source. Others remove misfolded or denatured proteins or are involved in the maturation of pre-proteins, the removal of signal peptides, and in the regulatory proteolysis of signal transduction proteins or transcriptional regulators. Furthermore, in pathogens, proteases are also important virulence factors. Reflecting this multiplicity of functions, in bacteria typically a broad set of proteases can be found. The composition of this set of proteolytic enzymes is specific for a single bacterium and depends on the ecological niche inhabited.

This review focuses on the role of proteases in corynebacteria. The proteolytic equipment of the sequenced *Corynebacterium* species is compared to the number of proteases in other important groups of Gram-positive bacteria, namely bacilli, streptomycetes, and mycobacteria. Recent data on the *Corynebacterium glutamicum* Clp and FtsH protease complexes are summarized and the putative role of secreted and surface-anchored proteases in pathogenic corynebacteria, especially in *Corynebacterium diphtheriae*, is discussed.

1. General functions of proteases
Proteolytic enzymes catalyze the cleavage of peptide bonds and represent approximately two percent of the total number of proteins in all types of organisms (Rao et al., 1998). This means that even in bacteria with strong reduction of genome size, such as in *Mycoplasma* species, several proteases can be found (a topical list of peptidases in various organisms is presented at the MEROPS database (Rawlings et al., 2006)). The multiplicity of these enzymes in cells is caused by the fact that proteases differ significantly in respect to their physiological function. For example, proteases can catalyze the total degradation of their substrate. Resulting amino acids or small peptides can subse-

296 Proteolysis in corynebacteria

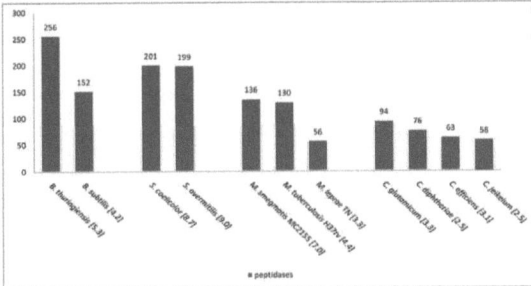

Figure 13.1. Peptidases and their homologues in selected Gram-positive bacteria. Shown are the total numbers of predicted putative peptidases according to the MEROPS database (Rawlings et al., 2006). Numbers in square brackets give the respective genome size in Mbp.

quently be used as building blocks for new proteins or as carbon and energy source. In this case, typically a set of rather unspecific endo- and exoproteases work together to ensure a fast and complete hydrolysis of their targets. Proteins used as growth substrates cannot be transported into a bacterial cell. Therefore, proteases are often transported across the cytoplasmic membrane and either bound to the surface of the cell or released into the surrounding medium. The resulting hydrolysis products are subsequently transported into the cell by specific amino acid or peptide uptake systems. Other proteases have a housekeeping function and degrade misfolded proteins or proteins damaged by heat, radiation or other detrimental factors (Wickner et al., 1999; Hengge and Bukau, 2003). Accessory subunits help to identify target proteins and to prevent hydrolysis of functional proteins in this case. Highly specific proteases are also involved in the maturation of preproteins (van Roosmalen et al., 2004) and in the regulatory proteolysis of signal transduction proteins or transcriptional regulators (Gottesman, 1999). Furthermore, in pathogens, proteases are involved in virulence and can damage host cells and tissues (Travis et al., 1995; Armstrong, 2006; Ribeiro-Guimaraes et al., 2007).

2. Distribution of proteases in bacilli, streptomycetes, and mycobacteria

Due to the multiple functions of proteases, all bacteria have a broad spectrum of these enzymes (for a topical list of peptidases in various species see MEROPS database (Rawlings et al., 2006)). Interestingly, the proteolytic capacity differs significantly between different bacteria. When major groups of Gram-positive bacteria are compared, namely bacilli, streptomycetes, mycobacteria, and corynebacteria (Figure 13.1), it becomes obvious that the spore-forming *Bacillus* species are proteolytically the most active group, especially when the number of proteases is correlated with genome size. Forty eight protease-encoding genes were annotated per Mbp of *Bacillus thuringiensis* chromosomal DNA and 36 genes per Mbp in the genome of *Bacillus subtilis*. The high number of proteins with putative proteolytic functions in *B. thuringiensis* might be the result of its pathogenic lifestyle. This bacterium kills insect larvae by a toxin, which destroys the gut epithel of the larvae, and uses the decaying bodies as nutrient source (Crickmore, 2005; Bravo et al., 2007).

Due to their high proteolysis activity and ability to secrete proteolytic enzymes into the surrounding medium, species like *B. subtilis, Bacillus licheniformis* or *Bacillus halodurans* are important industrial producers of alkaliphilic proteases as additives for laundry detergents, amongst various other secreted enzymes (Schallmey et al., 2004; Westers et al., 2004).

Despite their large genome sizes of about 9 Mbps, streptomycetes possess only a moderate number of proteases. In the *Streptomyces coelicolor* genome approximately 23 protease-encoding genes were found per Mbp chromosomal DNA, in case of *Streptomyces avermitilis* 22 genes. Taking the complex life cycle of streptomycetes and their various metabolic capabilities into account this is surprising. Polypeptides seem to be a less preferred growth substrate of these saprophytic soil bacteria, as streptomycetes can use various other growth substrates including the abundant polysaccharides chitin, xylan, and cellulose (Schrempf, 2001; Bertram et al., 2004).

Mycobacteria are also only moderately proteolytically active, especially *Mycobacterium smegmatis*, which features a comparatively small amount of proteins with peptidase function in respect to its genome size of about 7 Mbp. In this organism, only approximately 19 protease-encoding genes were annotated per Mbp of chromosomal DNA. An even more reduced number of proteases was found in *Mycobacterium leprae* with only 17 proteases encoded per Mbp. The reason might be the lifestyle of this obligate intracellular parasite resulting in severe gene decay termed *reductive evolution* (Eiglmeier et al., 2001). Nevertheless, some of the genes coding for proteins with putative proteolytic activity contribute to the process of human infection and leprosy skin lesions (Ribeiro-Guimaraes et al., 2007). This is also the case for the major human pathogen of the same genus, *Mycobacterium tuberculosis*, where proteases are studied as important pharmaceutical research targets in concerning infection and virulence (Ribeiro-Guimaraes and Pessolani, 2007). *M. tuberculosis* has the highest number of genes encoding proteases in this genus and approximately 30 of these genes are encoded per Mbp of chromosomal DNA (Rawlings et al., 2006), indicating a more important role of this enzyme class for survival of this bacterium.

3. The repertoire of protease-coding genes in corynebacteria

Based on published genome sequences of the *Corynebacterium* species *C. diphtheriae* (Cerdeno-Tarraga et al., 2003), *Corynebacterium efficiens* (Fudou et al., 2002), *Corynebacterium glutamicum* (Ikeda and Nakagawa,

Table 13.1. Protease-coding genes in sequenced genomes of corynebacteria. Amino acid sequence similarity searches were carried out using the BLAST algorithm (Altschul et al., 1990). The gene products were considered homologous when their identity was 50% or more and when the corresponding genes were situated in similar regions on the respective chromosomes. The putative proteases were validated based on protein domain searches according to the prediction programs in the Pfam (Bateman et al., 2004), MEROPS (Rawlings et al., 2006), and SMART (Letunic et al., 2006) databases. Non-classifiable secreted proteins with putative proteolytic functions are shown at the end of the table. Some of these proteins were already identified in proteomic studies (**bold**) but not yet further characterized. The obtained dataset was compared to proteome maps of the respective organism (Schaffer and Burkovski, 2005; Hansmeier et al., 2006a, 2006b, 2007; Lüdke et al., 2007; Lüdke, unpublished observations).

Protein name / family	*C. glutamicum* ATCC 13032 Bielefeld	*C. diphtheriae* NCTC13129	*C. efficiens* YS-314	*C. jeikeium* K411	Function
Signal peptidase I (SPase I, LepB)	*Cg2232*	*DIP1516*	*Ce1926*	*JK1183*	cleavage of hydrophobic, N-terminal signal or leader sequences from secreted proteins
Signal peptidase II (SPase II, LspA)	*Cg2347*	*DIP1584*	*Ce2033*	*JK0765*	release of signal peptides from bacterial membrane prolipoproteins including murein prolipoprotein.
Prepilin peptidase (PppA)	*Cg1830*	*DIP1346*	*Ce1744*	*JK1034*	processing of type 4 pilin precursor proteins to their mature forms by removal of leader peptides
	-	*DIP1621*	**Ce2080**	*JK0725*	
Cell wall associated and secreted proteases (NlpC/P60)	*Cg2401*	*DIP1622*	*Ce2081*	*JK0726*	probably peptidoglycan-lytic and/or invasion-associated
	Cg2402	-	*Ce1582*	-	
	Cg1636	*DIP0640*	*Ce0701*	*JK1678*	
	Cg0784	*DIP1281*	*Ce1659*	*JK0967*	
	Cg1735				

SrtA	-	-	-	JK1700	
SrtB	-	DIP2012	Cc2454	-	
SrtC	-	DIP0233	Cc2456	-	
SrtD	-	**DIP0236**	Cc2738	-	anchoring of proteins (e.g. pili proteins) at the outer face of the cell wall
SrtE	-	DIP2225	Cc2741	-	
SrtF	Cg3251	DIP2224	Cc2786	JK0103	
FtsH	**Cg2984**	DIP2272	Cc2542	JK0278	for further details, see text
HtpX	-	DIP2002	-	JK1570	heat shock protein; clostridial origin
ClpX	Cg2620	-	Cc2291	JK0546	
ClpP2	**Cg2644**	DIP1789	**Cc2311**	**JK0545**	
ClpP1	**Cg2645**	DIP1791	**Cc2312**	JK0544	for further details, see text and Figure 13.2
ClpC	**Cg2963**	DIP1792	Cc2529	JK0299	
ClpB	**Cg3079**	DIP1983	**Cc2613**	**JK0201**	
HtrA	Cg0998	DIP2104	Cc0950	**JK1530**	heat shock protein
Aminopeptidase II	Cg3325	DIP0856	Cc2832	-	putative ATP-dependent Zn-protease (*Bacillus*-type)
M50 membrane-associated protease	Cg2207	**DIP2346**	Cc1904	JK1166	probably involved in the regulation of gene expression by proteolysis of transcription regulators
protease II	**Cg2873**	DIP1499	Cc2486	JK0351	prolyl oligopeptidase B
subtilisin-like serine protease	Cg0665	DIP1926	Cc0578	JK1752	similar to mycosins, cell wall-associated
prenyl protease	Cg3209	DIP0554	Cc2734	JK0117	unknown
putative metalloprotease	Cg0029	DIP2248	Cc0316	-	unknown, probable chaperone-function, membrane bound

Table 13.1 continued

Protein name / family	C. glutamicum ATCC 13032 Bielefeld	C. diphtheriae NCTC13129	C. efficiens YS-314	C. jeikeium K411	Function
chymotrypsin-like proteases, secreted and precursors	Cg0356	DIP0307	Ce0292	**JK1968**	
	-	**DIP0350**	-	-	
	Cg0901	DIP0736	Ce0808	JK2027	
	Cg1243	DIP0743	Ce1149	JK1566	
	-	**DIP0964**	Ce1931	-	
PepO	Cg0193	DIP0154	Ce0152	JK0081	exo-/endo-aminopeptidases
PepE	**Cg1681**	DIP1238	Ce1613	**JK1241**	
PepC	**Cg1693**	**DIP1250**	Ce1628	-	
PepQ	**Cg1826**	DIP1341	Ce1738	JK1029	
PepB	Cg2419	DIP1637	Ce2096	JK0709	
PepN	Cg2662	DIP1798	Ce2320	JK0539	
PepA	-	-	-	JK1387	
AmiA	Cg0789	DIP0644	Ce0707	JK1763	peptidoglycan hydrolases
AmiC	Cg2285	DIP1551	Ce1986	JK0500	
AmiB	**Cg3021**	DIP2037	Ce2595	JK0802	
MapA	Cg0649	DIP0542	Ce0566	JK1764	methionyl aminopeptidases
MapB	Cg2198	DIP1496	Ce1901	JK1159	
SppA	Cg2043	-	-	-	signal peptide peptidase homolog (prophage encoded)

115

Zn-dependent metallopeptidase	Cg1842	**DIP1354**	Ce1753	JK1042	
	Cg0893	DIP0730	Ce0799	JK1605	
	Cg2902	-	**Ce2500**	-	
	Cg0349	DIP0302	Ce0286	JK1973	
	Cg0684	-	Ce0596	-	
	Cg1332	-	**Ce1275**	-	
	Cg3424	DIP2375	**Ce2935**	JK2095	
	Cg3131	DIP0104	-	-	
probable secreted proteases	Cg0409	DIP0363	Ce0353	JK1938	not further characterized ORFs with probable proteolytic functional domains
	Cg0410	DIP0364	Ce0354	-	
	Cg0650	**DIP0544**	**Ce0567**	**JK0340**	
	Cg0611	-	**Ce0536**	-	
	-	-	**Ce2413**	-	
	Cg1577	-	**Ce1522**	-	
	Cg0980	DIP0836	**Ce0934**	**JK1559**	
	Cg0388	-	Ce0331	-	
	-	-	Ce2742	-	

2003; Kalinowski et al., 2003) and *Corynebacterium jeikeium* (Tauch et al., 2005), a comparative genomics analysis of genes encoding proteins with putative proteolytic functions was performed. A complete overview of protease genes found in each of the corynebacterial genomes is given in Table 13.1. A total number of 53, 51, 56, and 42 protease-encoding genes was predicted in the genomes of *C. glutamicum*, *C. diphtheriae*, *C. efficiens*, and *C. jeikeium*. Correlated to the genome size, *C. efficiens* has the lowest density of protease-encoding genes on the chromosome (less than 20 per Mbp), followed by *C. jeikeium* (approximately 23 genes per Mbp), *C. glutamicum* (about 29 genes per Mbp) and *C. diphtheriae* (more than 30 genes per Mbp). A core subset of 39 homologous proteases is common to all corynebacterial species. These findings are in good accordance with the numbers recently published for the closely related mycobacterial genus (Ribeiro-Guimaraes et al., 2007), comprising a similar set of highly conserved homologous proteases.

Two of the core proteases that are found in all corynebacteria are the signal peptidases I and II that cleave signal peptides from secreted (lipo-)proteins. Another conserved protein with similar function is the prepilin peptidase that processes type 4 pilin precursor proteins (prepilins) to their mature forms by removal of leader peptides. A gene encoding the prepilin peptidase (family A24 peptidase) is present in all corynebacterial genomes, while *C. efficiens* features an additional family A24 peptidase of unknown function (Ce2562) for which no homologs exist in other corynebacteria.

Members of the NlpC/P60 protein family (Anantharaman and Aravind, 2003), probably invasion-associated proteins, are highly conserved among non-pathogenic and pathogenic corynebacteria. Genes encoding these putative proteases are found in all corynebacterial genomes, where they are always located in a pair wise manner directly adjacent to another (*Cg2401/2402*, *DIP1621/1622*, *JK0725/0726*, *Ce2080/2081*) possibly as a result of gene duplication in an early corynebacterial ancestor cell. Other conserved NlpC/P60 family proteins found in all species are the uncharacterized Cg0784/DIP0640/Ce0701/JK1678 proteins as well as another hypothetical invasion-associated protein (Cg1735, DIP1281, JK0967, Ce1659). Additionally, both, *C. glutamicum* and *C. efficiens* feature an unknown NlpC/P60 protein with putative proteolytic function, Cg1636 and Ce1582, for which no homologs in other actinomycetes are found.

In all corynebacterial genomes a highly conserved gene coding for sortase F (*Cg3251, DIP2272, JK0103, Ce2786*) was observed. SrtF homologs seem to be the house-keeping sortases of corynebacteria and homologs are present in most actinomycetes. Sortases covalently attach a variety of proteins to the exterior of the bacterial cell wall and contribute to the diseases caused by the pathogenic bacteria. While Cg3251 is the only sortase-like protein in *C. glutamicum*, in *C. jeikeium* additionally a *srtA* homolog, *JK1700*, is found. An even more complex set of *srt* genes was annotated in the genome sequences of *C. diphtheriae*, with six and *C. efficiens* with five corresponding gene clusters (Scott and Zähner, 2006; Marraffini et al., 2006). These pili-specific sortases seem to form their own subfamily and the corresponding genes are

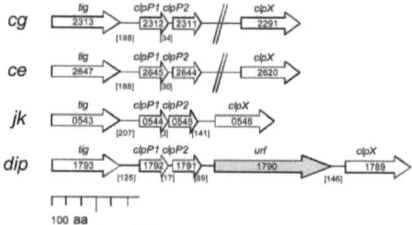

Figure 13.2. Genetic organization of *clp* gene clusters in corynebacteria. Arrows indicate the lengths and transcriptional orientations of annotated genes and predicted ORFs. Genes are shown by their annotation number, with the prefix of the respective organism. The gene names are assigned according to the annotations given by TIGR (http://www.tigr.org). Numbers in square brackets refer to the intergenic distance between two genes. Gene designations are as follows: *urf*, unknown reading frame; *tig*, trigger factor; *clpP1*, *clpP2*, proteolytic subunits; *clpX*, ATPase subunit; *cg*, *C. glutamicum*; *ce*, *C. efficiens*; *jk*, *C. jeikeium*; *dip*, *C. diphtheriae*.

located in clusters together with their respective pili-encoding genes (*spa*). Homologs of genes encoding the highly conserved membrane-bound FtsH protein are again found in all corynebacterial genomes (*Cg2984*, *Ce2542*, *DIP2002*, *JK0278*). The function of this AAA+ protein (ATPases associated with diverse cellular activities plus) in *C. glutamicum*, which is the only experimentally characterized FtsH protein in corynebacteria (Lüdke et al., 2007), will be discussed below. Also Clp protein complexes belong to the AAA+ class of proteases, which often have regulatory functions. These proteases are composed of accessory ATP-hydrolyzing subunits for recognition and unfolding of target proteins and proteins building a proteolytic chamber. The genes coding for ClpP1 and ClpP2 peptidase subunits are highly conserved in all available corynebacterial genomes. They are localized pair wise on the genome and are preceded by the trigger factor-encoding *tig* gene. A gene encoding the ATP-binding subunit ClpX is usually located in close neighbourhood on the chromosome (Figure 13.2), while the localization of the *clpC* gene, which encodes a second ATP-hydrolysing subunit, is less conserved. Details on targets and regulation of the Clp system are discussed below.

In the genus corynebacteria, only *C. jeikeium* features an ORF (*JK1570*) encoding a protein with the conserved catalytic domain of the *E. coli* HtpX endopeptidase that seems to be of clostridial origin and for which no other actinobacterial homologs are found. A protein with a putative role in heat shock response and probably involved in the degradation of mis- or unfolded proteins is encoded by the *htrA* homologs *Cg0998*, *DIP0856*, *Ce0950*, and *JK1530*.

Furthermore, an ORF encoding a putative M50 membrane-associated protease is found in all corynebacterial genomes. This not further characterized protease is highly conserved across all actinobacteria and seems to be crucial for this group of bacteria, deduced from the fact that a functional homolog is present even in *M. leprae*.

All corynebacterial genomes feature a diverse set of chymotrypsin-like proteases, of which some, like DIP0350 and Ce1931, seem to be unique to their respective organism. Most of them are synthesized as precursors and occur periplasmic or secreted with unknown functions. Common to all corynebacteria is also a more or less conserved set of various endo- and exo-aminopeptidases and carboxypeptidases that show a broad spectrum of putative biological functions, e. g. processing of proteins to mature forms, metabolism and lysis of bacterial cell walls, or assisting in the complete degradation of polypeptides to free amino acids.

Interestingly, the only corynebacterium to feature a signal peptide peptidase SppA homolog is *C. glutamicum*: Cg2043 is encoding a unique putative protease IV that seems to be of archaeobacterial origin, and for which no homolog is present in other corynebacterial or actinobacterial genomes; homologs of the actinobacterial SppA (e.g. *M. tuberculosis* Rv0724) are not found in any corynebacterial genomes.

4. AAA+ proteases and the *C. glutamicum* Clp system

Since hydrolysis of a peptide bond is an exergonic reaction (ΔG approx. -10 kJ mol^{-1}), no energy-rich cofactors such as ATP are necessary for proteolysis (Goldberg and St. John, 1976). Nevertheless, some proteases are ATP-dependent. These typically oligomeric protease complexes are characterized by a proteolytic chamber with a narrow entrance suitable only for small (unfolded) proteins. The ATPase domain located at this entrance is responsible for substrate recognition, unfolding, and transport into the proteolytic core (Hlavacek and Vachova, 2002). All ATP-dependent proteases described so far belong to the AAA+ superfamily (Ogura and Wilkinson, 2001; Lupas and Martin, 2002). Five different AAA+ proteases were described in prokaryotes, namely, the proteasome, HslUV, ClpAP (and its homolog in Gram-positive bacteria, ClpCP), ClpXP, and FtsH. In corynebacteria, genes encoding the proteasome or HslUV were not detected; Clp and FtsH proteins are discussed in the following paragraphs.

In *C. glutamicum*, the Clp proteins are the best investigated proteolytic enzymes. Typically for actinomycetes, two genes, *clpP1* and *clpP2*, are located adjacent to each other (Figure 13.2), encoding the subunits of the proteolytic core. ClpC and ClpX are accessory ATPase subunits, which are members of the Clp/Hsp100 superfamily (Schirmer et al., 1996). Additionally, a *clpB* gene is annotated in the genome sequence. The corresponding protein has chaperone function and is not part of the Clp protease complexes.

A first hint for Clp function was found when the heat shock response of *C. glutamicum* was investigated. As shown by transcriptome analyses, expression of *clpP1*, *clpP2*, and *clpC* is increased in response to a heat shock (Muffler et al., 2002). Part of the underlying regulatory mechanism was already elucidated (Engels et al., 2004; 2005). Proteome analyses of

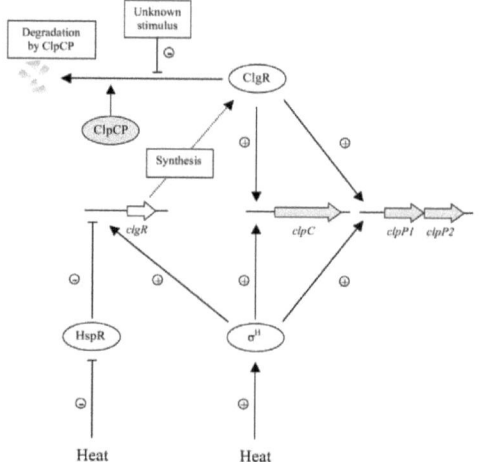

Figure 13.3. Regulation of *clpC* and *clpP1P2* transcription in *C. glutamicum*. Expression of *clpC* and *clpP1P2* genes is controlled by the transcriptional regulators HspR and ClgR. Additionally, expression is influenced by the ECF sigma factor σ^H. In response to heat shock, σ^H leads to activation of transcription *via* complex formation with RNA polymerase (RNAP). Besides direct induction of *clpC* and *clpP1P2* transcription by σ^H-RNAP, additionally σ^H-RNAP activates also the expression of the transcriptional activator ClgR. The resulting increase in ClgR concentration does only result in an increase of *clpC* and *clpP1P2* expression, when ClpC-dependent degradation of ClgR is blocked by an unknown stimulus. Furthermore, heat shock inactivates the transcriptional repressor HspR (Engels et al., 2004).

a *clpC* deletion strain in comparison to the wild type showed an approximately tenfold increased abundance of ClpP1 and ClpP2 in the mutant; the mRNA levels of the corresponding genes were increased by a factor of about five (Engels et al., 2004). DNA affinity purification experiments using the *clpP1P2* promoter region as bait resulted in the identification of ClgR. Further analyses showed that transcription of *clpP1*, *clpP2*, and *clpC* is controlled not only by this transcriptional regulator, but by a network of different regulatory proteins including ClgR, HspR, and the ECF (extracytoplasmic function) sigma factor σ^H (Figure 13.3). The regulon of ClgR was investigated in more detail by transcriptome profiling, primer extensions, and bioinformatic analyses, which revealed that at least five genes, namely *NCgl0240*, *NCgl0748*, *hflX*, *ptrB*, and *recR* are regulated by ClgR, while for about ten other genes a direct or indirect regulation by ClgR is discussed. When the (annotated) functions of the corresponding gene products are taken in consideration, it becomes obvious that this transcriptional activator controls expression of genes encoding proteins involved in proteolysis and DNA repair (Engels et al., 2005).

In an independent study it was shown that *C. glutamicum clpC* and *clpX* deletion mutations as well as a conditional *clpP1P2* mutant strain revealed an impaired degradation of the nitrogen signal transduction protein GlnK (Strösser et al., 2004). Since membrane contact is crucial for GlnK degradation, an indirect effect of the cytoplasmic Clp protease complexes on GlnK proteolysis was favoured. However, using a ClpP-specific polyclonal antiserum, also a minor fraction of Clp proteins was detectable at the membrane leaving the question of an involvement of Clp proteases in GlnK degradation open (Strösser et al., 2004).

5. The role of FtsH in *C. glutamicum*

FtsH proteins are membrane anchored proteases and were described as "machines for degrading membrane proteins" (Langer, 2000). They belong as Clp proteins to the AAA+ group of proteases, which often have regulatory functions. In fact, for the *C. glutamicum* FtsH an effect on the degradation of nitrogen signal control protein GlnK was reported (Strösser et al., 2004). The membrane localization of FtsH hints to a direct function on GlnK, since GlnK is only degraded, when it is sequestered to the membrane (Strösser et al., 2004). However, experimental evidence for an interaction of *C. glutamicum* FtsH and GlnK resulting in proteolysis of GlnK by FtsH is still missing.

Mutations of *ftsH* were described in different bacteria and were found to be remarkably species-specific. Effects of *ftsH* mutations range from drastic growth impairment in *Escherichia coli* (Begg et al., 1992) to effects on sporulation, development and stress response in *B. subtilis* (Deuerling et al., 1997; Lysenko et al., 1997) and *Caulobacter crescentus* (Fischer et al., 2002). In *C. glutamicum* cells, a deletion of the *ftsH* gene is well tolerated and has, in contrast to the situation in other bacteria, no strong effect on growth. Only a slightly increased doubling time of 2 h 38 min for the *ftsH* mutant compared to 2 h 18 min in the wild type was observed in minimal medium at 30 °C (Lüdke et al., 2007). This effect was more pronounced at 37 °C (Lüdke, unpublished observations), while no effect was found when cells were challenged by a sudden temperature increase to 39 °C or sudden or constant osmotic stress (Lüdke et al., 2007). Obviously, FtsH plays a less crucial role in *C. glutamicum* compared to other bacteria.

Besides by growth experiments, the influence of the membrane-bound AAA+ protease FtsH on membrane and cytoplasmic proteins of *C. glutamicum* was investigated in more detail. Proteome analyses using a combination of anion exchange chromatography and SDS-PAGE for the membrane fraction and standard two-dimensional gel electrophoresis for cytoplasmic proteins were carried out to identify FtsH targets in the cell (Lüdke et al., 2007). It was shown that deletion of the *ftsH* gene causes a strong increase of nine cytoplasmic and membrane proteins, namely biotin carboxylase/biotin carboxyl carrier protein, glyceraldehyde-3-phosphate dehydrogenase, homocysteine methyltransferase, malate synthase, isocitrate lyase, a conserved hypothetical protein (NCgl1985), succinate dehydrogenase A, succinate dehydrogenase B, and glutamate binding protein GluB. Thirty eight cytoplasmic and membrane-associated proteins showed a decreased abundance in an FtsH-lacking strain compared to the wild type. The decreasing amount of succinate dehydrogenase A in the cytoplasmic fraction of the *ftsH* mutant compared to the wild type and its increasing abundance in the membrane fraction indicates that FtsH might be involved in the cleavage of a membrane anchor of this membrane-associated protein and by this changes the localization of this enzyme. In summary, an astonishingly small amount of membrane and cytoplasmic proteins is affected by an *ftsH* deletion and the proteome analyses hint to a function of FtsH in regulation of energy and carbon metabolism as well as amino acid biosynthesis. A major role in stress response, as observed for other bacteria, was not found.

6. Proteases as virulence factors in corynebacteria

As mentioned above, proteases are not only involved in housekeeping functions and regulations, but also play an important role as virulence factors in pathogenic bacteria such as *Streptococcus pneumoniae*, *Pseudomonas aeruginosa*, and *Staphylococcus aureus* (Bethe et al., 2001; Malloy et al., 2005; Rigoulay et al., 2005; for review, see Armstrong, 2006). Often these host cell-damaging proteases are bound to the surface of the pathogen or secreted directly into the surrounding medium.

For *C. diphtheriae* and *C. jeikeium* detailed proteome analyses of surface-anchored and extracellular proteins were carried out recently (Hansmeier et al., 2006a; 2007). Interestingly, less proteases were identified in the nosocomial pathogen *C. jeikeium* compared to *C. diphtheriae*. The lipophilic lifestyle of *C. jeikeium* seems to reduce the need for a high proteolytic capacity. In contrast, several different proteases were identified that are located at the cell surface of *C. diphtheriae* or are secreted into the medium (Hansmeier et al., 2006a; see also Tab. 1). Initiated by these observations, the proteolytic capacity of *C. diphtheriae* was tested in respect to haemolysis of blood agar and halo generation on milk plates. Compared to *Bacillus cereus*, a negligible haemolysis and protease activity was observed (A. Wünsche, unpublished observation). In order to characterize major secreted proteases in more detail, mutants were generated, lacking the trypsin-like serine proteases DIP0350 and DIP0964, respectively. The gene encoding DIP0350 is localized on a genome island (see Genomics chapter) and was discussed to be involved in pathogenicity of *C. diphtheriae* (Hansmeier et al., 2006a). The two genes are expressed in a growth-dependent manner with the highest level of mRNA found in the exponential growth phase. Interestingly, loss of these proteases in corresponding mutant strains led to a significantly impaired ability to bind to host cells *in vitro* (A. Wünsche, unpublished observation). These observations indicate that proteolytic degradation of epithel tissue is not a major strategy of *C. diphtheriae* to colonize its host and proteases might rather be directly or indirectly involved in adhesion to epithelial cells.

The function of other proteolytic enzymes on the adherence of *C. diphtheriae* to host cells was demonstrated recently in a study addressing the function of sortases (Mandlik et al., 2007). As shown in Table 13.1, six different sortase-encoding genes are annotated in the *C. diphtheriae* genome sequence (Pallen et al., 2001; Cerdeno-Tarraga et al., 2003). The *srtA-E* genes are localized in three different clusters together with the pilus assembly genes *spaA-I*. While sortase SrtA is essential for the polymerization of SpaABC pili (Ton-That and Schneewind, 2003), SpaDEF pili require SrtB and SrtC (Gaspar and Ton-That, 2006), and the SrtD and SrtE sortases are required for the assembly of SpaHIG pili (Swierczynski and Ton-That, 2006). A *C. diphtheriae* mutant lacking all sortases, which is unable to assemble any proper pilus structures, showed a dramatically reduced binding to pharynx cells (Mandlik et al., 2007), emphasizing the importance of proteolysis in respect to pathogenicity in this organism.

7. Concluding remarks

Compared to other bacteria, corynebacteria seem to have only moderate proteolytic activity. Nevertheless, proteases obviously play a crucial role in many cellular processes. First investigations of *C. glutamicum* Clp protease complexes and FtsH hint to interesting regulatory functions of these AAA+ enzymes in this species. However, knowledge on this class of proteases is far from completeness and for example even the degradation signals are still unknown. The same is true for the function of putative degradation tags or amino acid residues influencing protein stability (Mogk et al., 2007). As discussed above, already the increased density of protease-encoding genes in the *C. diphtheriae* genome compared to other corynebacterial genomes and their partial localization on genomic islands suggest a role of these enzymes in pathogenicity. In fact, *C. diphtheriae* sortases were already shown to be crucial for pili assembly and adhesion to hosts cells, an effect also observed for two other secreted *C. diphtheriae* proteases. However, also in this case, only limited information is available. Future approaches might shed more light on the diverse function of known and uncharacterized proteases in different corynebacteria and hopefully elucidate the up to now completely unknown target recognition mechanisms of these enzymes.

References

Altschul, S.F., Gish, W., Miller, W., Myers, E.W., Lipman, D.J. (1990). Basic local alignment search tool. J. Mol. Biol. *215*, 403-410.

Anantharaman, V., and Aravind, L. (2003). Evolutionary history, structural features and biochemical diversity of the NlpC/P60 superfamily of enzymes. Genome Biol. *4*, R11.

Armstrong, P. (2006). Proteases and protease inhibitors: a balance of activities in host-pathogen interaction. Immunobiology *211*, 263-281.

Bateman, A., Coin, L., Durbin, R., Finn, R.D., Hollich, V., Griffiths-Jones, S., Khanna, A., Marshall, M., Moxon, S., Sonnhammer, E.L.L., Studholme, D.J., Yeats, C., and Eddy, S.R. (2004). The Pfam protein families database. Nucleic Acids Res. *32*, D138-D141.

Begg, K.J., Tomoyasu, T., Donachie, W.D., Khattar, M., Niki, H., Yamanaka, K., Hiraga, S., and Ogura, T. (1992). *Escherichia coli* mutantY16 is a double mutant carrying thermosensitive *ftsH* and *ftsI* mutations. J. Bacteriol. *174*, 2516-2417.

Bertram, R., Schlicht, M., Mahr, K., Nothaft, H., Saier, M.H., and Titgemeyer, F. (2004). *In silico* and transcriptional analysis of carbohydrate uptake systems of *Streptomyces coelicolor* A3(2). J. Bacteriol. *186*, 1362-1373.

Bethe, G., Nau, R., Wellmer, A., Hakenbeck, R., Reinert, R.R., Heinz, H.-P., and Zysk, G. (2001). The cell wall-associated serine protease PrtA: a highly conserved virulence factor of *Streptococcus pneumoniae*. FEMS Microbiol. Lett. *205*, 99-104.

Bravo, A., Gill, S.S., and Soberon, M. (2007). Mode of action of *Bacillus thuringiensis* Cry and Cyt toxins and their potential for insect control. Toxicon *49*, 423-435.

Cerdeno-Tarraga, A.M., Efstratiou, A., Dover, L.G., Holden, M.T.G., Pallen, M., Bentley, S.D., Besra, G.S., Churcher, C., James, K.D., De Zoysa, A., Chillingworth, T., Cronin, A., Dowd, L., Feltwell, T., Hamlin, N., Holroyd, S., Jagels, K., Moule, S., Quail, M.A., Rabbinowitch, E., Rutherford, K.M., Thomson, N.R., Unwin, L., Whitehead, S., Barrell, B.G., and Parkhill, J. (2003). The complete genome sequence and analysis of *Corynebacterium diphtheriae* NCTC13129. Nucleic Acids Res. *31*, 6516-6523.

Crickmore, N. (2005). Using worms to better understand how *Bacillus thuringiensis* kills insects. Trends Microbiol. *13*, 347-350.

Deuerling, E., Mogk, A., Richter, C., Purucker, M., and Schumann, W. (1997). The *ftsH* gene of *Bacillus subtilis* is involved in major cellular processes such as sporulation, stress adaptation and secretion. Mol. Microbiol. *23*, 921-933.

Eiglmeier, K., Parkhill, J., Honore, N., Garnier, T., Tekaia, F., Telenti, A., Klatser, P., James, K.D., Thomson, N.R., Wheeler, P.R., Churcher, C., Harris, D., Mungall, K., Barrell, B.G., and Cole, S.T. (2001). The decaying genome of *Mycobacterium leprae*. Lepr. Rev. *72*, 387-398.

Engels, S., Schweitzer, J.-E., Ludwig, C., Bott, M., and Schaffer, S. (2004). *clpC* and *clpP1P2* gene expression in *Corynebacterium glutamicum* is controlled by a regulatory network involving the transcriptional regulators ClgR and HspR as well as the ECF sigma factor σ^H. Mol. Microbiol. *52*, 285-302.

Engels, S., Ludwig, C., Schweitzer, J.-E., Mack, C., Bott, M., and Schaffer, S. (2005). The transcriptional regulator ClgR controls transcription of genes involved in proteolysis and DNA repair in *Corynebacterium glutamicum*. Mol. Microbiol. *57*, 576-591.

Fischer, B., Rummel, G., Aldrige, P., and Jenal, U. (2002). The FtsH protease is involved in development, stress response and heat shock control in *Caulobacter crescentus*. Mol. Microbiol. *44*, 461-478.

Fudou, R., Jojima, Y., Seto, A., Yamada, K., Rimura, E., Nakamatsu, T., Hirashi, A., and Yamanaka, S. (2002). *Corynebacterium efficiens* sp. Nov., a glutamic-acid-producing species from soil and plant material. Int. J. Syst. Evol. Microbiol. *52*, 1127-1131.

Gaspar, A.H., and Ton-That, H. (2006). Assembly of distinct pilus structures on the surface of *Corynebacterium diphtheriae*. J. Bacteriol. *188*, 1526-1533.

Goldberg, A.L., and St. John, A.C. (1976). Intracellular protein degradation in mammalian and bacterial cells – part 2. Ann. Rev. Biochem. *45*, 747-803.

Gottesman, S. (1999). Regulation by proteolysis: developmental switches. Curr. Opin. Microbiol. *2*, 142-147.

Hansmeier, N., Chao, T.C., Kalinowski, J., Pühler, A., and Tauch, A. (2006a). Mapping and comprehensive analysis of the extracellular and cell surface proteome of the human pathogen *Corynebacterium diphtheriae*. Proteomics *6*, 2465-2476.

Hansmeier, N., Chao, T.C., Pühler, A., and Tauch, A. (2006b). The cytosolic, cell surface and extracellular proteomes of the biotechnologically important soil bacterium *Corynebacterium efficiens* YS-314 in comparison to those of *Corynebacterium glutamicum* ATCC 13032. Proteomics *6*, 233-250.

Hansmeier, N., Chao, T.C., Daschkey, S., Müsken, M., Kalinowski, J., Pühler, A., and Tauch, A. (2007). A comprehensive proteome map of the lipid-requiring nosocomial pathogen *Corynebacterium jeikeium* K411. Proteomics *7*, 1076-1096.

Hengge, R., and Bukau, B. (2003). Proteolysis in prokaryotes: protein quality control and regulatory principles. Mol. Microbiol. *49*, 1451-1462.

Hlavacek, O., and Vachova, L. (2002). ATP-dependent proteinases in bacteria. Folia Microbiol. *47*, 203-212.

Ikeda, M., and Nakagawa, S. (2003). The *Corynebacterium glutamicum* genome: features and impacts on biotechnological processes. Appl. Microbiol. Biotechnol. *62*, 99-109.

Kalinowski, J., Bathe, B., Bischoff, N., Bott, M., Burkovski, A., Dusch, N., Eggeling, L., Eikmanns, B.J., Gaigalat, L., Goesmann, A., Hartmann, M., Huthmacher, K., Krämer, R., Linke, B., McHardy, A.C., Meyer, F., Möckel, B., Pfefferle, W., Pühler, A., Rey, D., Rückert, C., Sahm, H., Wendisch, V.F., Wiegräbe, I., and Tauch, A. (2003). The complete *Corynebacterium glutamicum* ATCC 13032 genome sequence and its impact on the production of L-aspartate-derived amino acids and vitamins. J. Biotechnol. *104*, 5-25.

Langer, T. (2000). AAA proteases: cellular machines for degrading membrane proteins. Trends Biol. Sci. *25*, 247-251.

Letunic, L., Copley, R.R., Pils, B., Pinkert, S., Schultz, J., and Bork, P. (2006). SMART 5: domains in the context of genomes and networks. Nucleic Acids Res. *34*, D257-D260.

Lüdke, A., Krämer, R., Burkovski, A., Schluesener, D., and Poetsch, A. (2007). A proteomic study of *Corynebacterium glutamicum* AAA+ protease FtsH. BMC Microbiol. *7*, 6.

Lupas, A.N., and Martin, J. (2002). AAA proteins. Curr. Opin. Struct. Biol. *12*, 746-753.

Lysenko, E., Ogura, T., and Cutting, S.M. (1997). Characterization of the *ftsH* gene of *Bacillus subtilis*. Microbiology *143*, 971-978.

Malloy, J.L., Veldhuizen, R.A.W., Thibodeaux, B.A., O'Callaghan, R.J., and Wright, J.R. (2005). *Pseudomonas aeruginosa* protease IV degrades surfactant proteins and inhibits surfactant host defense and biophysical functions. Am. J. Physiol. Lung Cell. Mol. Physiol. *288*, L409-L418.

Mandlik, A., Swierczynski, A., Das, A., and Ton-That, H. (2007). *Corynebacterium diphtheriae* employs specific minor pilins to target human pharyngeal epithelial cells. Mol. Microbiol. *64*, 111-124.

Marraffini, L.A., Dedent, A.C., Schneewind, O. (2006). Sortases and the art of anchoring proteins to the envelopes of gram-positive bacteria. Microbiol. Mol. Biol. Rev. *70*, 192-221.

Mogk, A., Schmidt, R., and Bukau, B. (2007). The N-end rule pathway for regulated proteolysis: prokaryotic and eukaryotic strategies. Trends Cell Biol. *17*, 165-172.

Muffler, A., Bettermann, S., Haushalter, M., Horlein, A., Neveling, U., Schramm, M., and Sorgenfrei, O. (2002). Genome-wide transcription profiling of

Corynebacterium glutamicum after heat shock and during growth on acetate and glucose. J. Biotechnol. *98*, 255-268.

Ogura, T., and Wilkinson, A.J. (2001). AAA+ superfamily ATPases: common structure – diverse function. Genes Cells *6*, 575-597.

Pallen, M.J., Lam, A.C., Antonio, M., and Dunbar, K. (2001). An embarrassment of sortases – a richness of substrates? Trends Microbiol. *9*, 97-102.

Rao, M.B., Tanksale, A.M., Ghatge, M.S.,and Deshpande, V.V. (1998). Molecular and biotechnological aspects of microbial proteases. Microbiol. Mol. Biol. Rev. *62*, 597-635.

Rawlings, N.D., Morton, F.R., and Barrett, A.J. (2006). MEROPS: the peptidase database. Nucleic Acids Res. *34*, D270-D272.

Ribeiro-Guimaraes, M.L., and Pessolani, M.C. (2007). Comparative genomics of mycobacterial proteases. Microb. Pathog., Epub ahead of print.

Ribeiro-Guimaraes, M.L., Tempone, A.J., Amaral, J.J., Nery, J.A., Antunes, S.L., and Pessolani, M.C. (2007). Expression analysis of proteases of *Mycobacterium leprae* in human skin lesions. Microb. Pathog., Epub ahead of print.

Rigoulay, C., Entenza, J.M., Halpern, D., Widmer, E., Moreillon, P., Poquet, I., and Gruss, A. (2005) Comparative analysis of the roles of HtrA-like surface proteases in two virulent *Staphylococcus aureus* strains. Infect. Immun. 73, 563-572.

Schaffer, S., and Burkovski, A. (2005). Genome-based approaches: proteomics. In: Handbook of *Corynebacterium glutamicum*. M. Bott, and L. Eggeling, eds. (CRC Press LLC, Boca Raton, FL.), pp. 99-118.

Schallmey, M., Singh, A., and Ward, O.P. (2004). Developments in the use of *Bacillus* species for industrial production. Can. J. Microbiol. *50*, 1-17.

Schirmer, E.C., Glover, J.R., Singer, M.A., and Lindquist, S. (1996). HSP100/Clp proteins: a common mechanism explains diverse functions. Trends Biochem. Sci. *21*, 289-296.

Schrempf, H. (2001). Recognition and degradation of chitin by streptomycetes. Antonie van Leeuwenhook 79, 285-289.

Scott, J.R., and Zähner, D. (2006). Pili with strong attachments: Gram-positive bacteria do it differently. Mol. Microbiol. *62*, 320-330.

Strösser, J., Lüdke, A., Schaffer, S., Krämer, R., and Burkovski, A. (2004). Regulation of GlnK activity: modification, membrane sequestration, and proteolysis as regulatory principles in the network of nitrogen control *in Corynebacterium glutamicum*. Mol. Microbiol. *54*, 132-147.

Swierczynski, A., and Ton-That, H. (2006). Type III pilus of corynebacteria: pilus length is determined by the level of its major pilin subunit. J. Bacteriol. *188*, 6318-6325.

Tauch, A., Kaiser, O., Hain, T., Goesmann, A., Weisshaar, B., Albersmeier, A., Bekel, T., Bischoff, N., Brune, I., Chakraborty, T., Kalinowski, J., Meyer, F., Rupp, O., Schneiker, S., Viehoever, P., and Pühler, A. (2005). Complete genome sequence and analysis of the multiresistant nosocomial pathogen *Corynebacterium jeikeium* K411, a lipid-requiring bacterium of the human skin flora. J. Bacteriol. *187*, 4671-4682.

Ton-That, H., and Schneewind (2003). Assembly of pili on the surface of *Corynebacterium diphtheriae*. Mol. Microbiol. *50*, 1429-1438.

Travis, J., Potempa, J., and Maeda, H. (1995). Are bacterial proteases pathogenic factors? Trends Microbiol. *3*, 405-407.

van Roosmalen, M.L., Geukens, N., Jongbloed, J.D.H., Tjalsma, H., Dubois, J.-Y.F., Bron, S., van Dijl, J.M., and Anné, J. (2004). Type I signal peptidases of Grampositive bacteria. Biochim. Biophys. Acta *1694*, 279-297.

Westers, L., Westers, H. and Quax, W.J. (2004). *Bacillus subtilis* as cell factory for pharmaceutical proteins: a biotechnological approach to optimize the host organism. Biochim. Biophys. Acta *1694*, 299-310.

Wickner, S., Maurizi, M.R., Gottesman, S. (1999). Posttranslational quality control: folding, refolding, and degrading proteins. Sience *286*, 1888-1893.

i want morebooks!

Buy your books fast and straightforward online - at one of world's fastest growing online book stores! Environmentally sound due to Print-on-Demand technologies.

Buy your books online at
www.get-morebooks.com

Kaufen Sie Ihre Bücher schnell und unkompliziert online – auf einer der am schnellsten wachsenden Buchhandelsplattformen weltweit! Dank Print-On-Demand umwelt- und ressourcenschonend produziert.

Bücher schneller online kaufen
www.morebooks.de

 VDM Verlagsservicegesellschaft mbH
Heinrich-Böcking-Str. 6-8
D - 66121 Saarbrücken

Telefon: +49 681 3720 174
Telefax: +49 681 3720 1749

info@vdm-vsg.de
www.vdm-vsg.de

Printed by Books on Demand GmbH, Norderstedt / Germany